Agricultural Extension

NIPA® GENX ELECTRONIC RESOURCES & SOLUTIONS P. LTD.
New Delhi-110 034

About the Author

Dr. P. Jaisridhar, is a Doctoral Degree holder in Agricultural Extension serving Tamil Nadu Agricultural University as an Assistant Professor in Agricultural Extension presently stationed at the Directorate of Extension Education, Coimbatore. With an extensive background, the author has previously worked at prestigious institutions including the Centre of Excellence in Millets, Tiruvannamalai, and Krishi Vigyan Kendra, Virinjipuram, Vellore.

His academic prowess is evident through his teaching portfolio, where he has conducted various courses for undergraduate and postgraduate programs, organized educational tours, and supervised thesis research. Additionally, Dr. Jaisridhar is a prolific author, having published books, contributed chapters, and presented papers at international and national conferences.

In the realm of research, Dr. Jaisridhar has made significant contributions, with several research papers published as the first author and leading externally funded schemes. His dedication to extension services is reflected in his organization of training programs, field demonstrations, and media engagements, contributing to agricultural development and knowledge dissemination.

Dr. Jaisridhar's institutional development initiatives highlight his proactive approach to enhancing organizational capabilities, including his involvement in student clubs and various administrative responsibilities. With a commitment to excellence, Dr. P. Jaisridhar continues to make substantial contributions to the field of agricultural extension and rural development.

Agricultural Extension Fundamentals and Facts

P. Jaisridhar Ph.D.
Assistant Professor (Agricultural Extension)
Directorate of Extension Education
Tamil Nadu Agricultural University
Coimbatore- 641 003

NIPA® GENX ELECTRONIC RESOURCES & SOLUTIONS P. LTD.
New Delhi-110 034

NIPA® GENX ELECTRONIC RESOURCES & SOLUTIONS P. LTD.

101,103, Vikas Surya Plaza, CU Block
L.S.C.Market, Pitam Pura, New Delhi-110 034
Ph : +91 11 4386 0225, 9717133558, 9540816132
E-mail: newindiapublishingagency@gmail.com
Website: www. niparesources.com

Print ISBN: 978-93-58878-94-3

ebook ISBN: 978-93-58874-36-5

Composed and Designed by NIPA®.

Preface

In the dynamic landscape of agricultural development, effective extension services are the lifeblood that nurtures growth and prosperity. "Agricultural Extension - Fundamentals & Facts" is an invaluable resource that illuminates the core principles, historical perspectives, and contemporary trends shaping the field of agricultural extension.

The journey begins with an exploration of education in its myriad forms, from formal institutions to informal and non-formal modes of learning. As we delve into the essence of extension education and its application in agricultural contexts, readers are introduced to foundational concepts, objectives, and guiding principles that underpin effective extension practices.

Rural development emerges as a central theme, with a comprehensive examination of its meaning, objectives, and the array of challenges it seeks to address. From the pioneering experiments of the pre-independence era to the transformative initiatives of the post-independence era, including community development programs and the evolution of the Panchayat Raj system, this book traces the evolution of rural development paradigms in India.

A detailed analysis of key agricultural development programs sheds light on their inception, objectives, and salient features, offering readers a nuanced understanding of their impact on agricultural productivity and rural livelihoods. Social justice and poverty alleviation programs, along with emerging trends such as privatization and the reorganized extension system, are also scrutinized, reflecting the evolving priorities and strategies in agricultural extension.

Special attention is devoted to women's development programs, recognizing the pivotal role of women in agricultural production and community empowerment. From the Development of Women and Children in Rural Areas (DWCRA) to the Mahila Samriddi Yojana (MSY), these initiatives underscore the imperative of gender-inclusive approaches in agricultural extension efforts.

The reorganized extension system, particularly the Training and Visit (T&V) system, is dissected to reveal its salient features, strengths, and limitations, paving the way for discussions on the emergence of Broad-Based Extension (BBE) models that strive for greater inclusivity and effectiveness.

"Agricultural Extension - Fundamentals & Facts" is not merely a textbook; it is a compendium of knowledge, insights, and practical wisdom garnered from decades of experience in the field. It is intended to serve as a beacon for extension practitioners, policymakers, researchers, and students alike, guiding them towards innovative approaches and sustainable solutions in agricultural development. As we embark on this journey together, may this book inspire transformative action and foster a brighter future for rural communities worldwide.

P. Jaisridhar

Contents

1

Education
Meaning, Definition & Types

Education is the intricate process of fostering an individual's capability to effectively respond to their circumstances. It encompasses the cultivation of desirable changes in human behavior, encompassing knowledge, attitudes, skills, values, and habits. This transformative process occurs autonomously and under the guidance of instructors, manifesting in various settings, formal or informal. Any experience that shapes one's cognition, emotions, or actions can be deemed educational.

Types of Education

1. Formal Education

Formal education denotes a structured, systematic model of learning, governed by established laws and standards, featuring a rigid curriculum concerning objectives, content, and methodology. It entails a continuous process of education, typically facilitated within educational institutions such as schools and universities. Key characteristics include:

- Planned with specific objectives.
- Confined within a defined timeframe.
- Follows a well-defined and structured curriculum.
- Administered by qualified educators.
- Extends beyond classroom settings.
- Upholds strict disciplinary measures.

2. Informal Education

Diverging from formal education, informal education is characterized by its lack of organization and systematic approach. It is a lifelong process wherein individuals accumulate knowledge, skills, attitudes, and insights from daily experiences and familial interactions. Salient features include:

- Incidental and spontaneous in nature.
- Not pre-planned.
- Not facilitated by specialized agencies.
- Lacks a prescribed timetable or curriculum.
- Can encompass negative influences as well.

3. Non-formal Education

Non-formal education comprises organized, systematic educational activities conducted outside the formal education system, targeting specific groups of the population. It serves to provide tailored learning experiences to address identified needs. Key attributes include:

- Derived from the concept of formal education.
- Operates outside the formal education structure.
- Intentional and deliberate in its approach.
- Tailored for homogeneous groups.
- Addresses the specific needs of the targeted population.

In summary, education manifests in diverse forms, each catering to distinct learning needs and contexts. Understanding the nuances of formal, informal, and non-formal education is essential for designing comprehensive educational strategies and fostering holistic development within societies.

2

Extension Education Meaning, Definition, Scope and Process

Extension Education

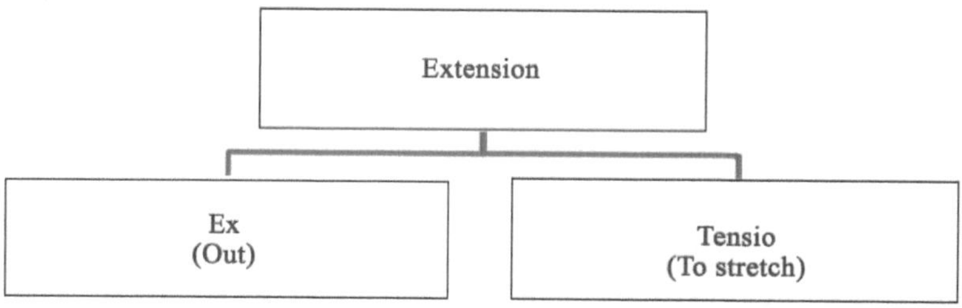

Extension education, stemming from the Latin roots ‘ex’, meaning out, and ‘tension’, signifying stretching, refers to an educational approach extended beyond the conventional boundaries of formal institutions, particularly targeting rural communities. Education serves as a pivotal component within extension, aiming to disseminate knowledge and foster development among diverse populations.

Origin and Evolution of Extension Education

The history of agricultural extension closely intertwines with technological advancements. Agricultural societies emerged in the 18th century, with Scotland hosting the first in 1728, dedicated to disseminating agricultural knowledge through lectures and publications. The term “University Extension” was introduced in Britain around 1840, focusing on making knowledge accessible to the general populace. James Stuart, a Fellow of Trinity College, Cambridge, pioneered practical extension efforts by delivering lectures to various associations in England in 1867-68. The concept gained formal recognition with Cambridge University’s adoption in 1873, followed by London University and Oxford University. In the United States, legislative acts such as the Morrill Act of 1862 and 1890, alongside the Hatch Act of 1887 and Smith-Lever Act of 1914, laid the groundwork for agricultural extension initiatives. India witnessed the establishment of the first agricultural university,

Govind Ballabh Pant University of Agriculture and Technology, Pantnagar, in 1960, emphasizing the significance of extension in agricultural development.

Growth of Agricultural Extension Education in India

The inception of extension education at the undergraduate level commenced in 1950 at the College of Agriculture, Calcutta University. Post-graduate programs followed suit, with Bihar Agricultural College, Sabour, launching the first in 1955, and the College of Nagpur in 1958. A pivotal milestone occurred with the establishment of a dedicated Division of Agricultural Extension at the Indian Agricultural Research Institute (IARI), New Delhi, with guidance from Dr. J. Paul Leagans. This initiative led to the introduction of Ph.D. programs in extension education at IARI in 1961, subsequently emulated by Punjab Agricultural University, Ludhiana.

Basic Definitions of Extension Education

Extension education encompasses an applied social science, integrating knowledge from physical, biological, and social sciences into a comprehensive framework aimed at providing non-credit education, predominantly targeting adults.

Ensminger (1957) underscores extension's purpose in altering attitudes and practices.

Leagans (1961) defines extension as out-of-school education aimed at enhancing farm, home, and community practices.

Leagans (1971) defines extension as a science facilitating planned changes in the behavior complex of individuals, aimed at enhancing their livelihoods.

Reddy (1993) emphasizes the role of extension in disseminating knowledge to improve vocational, entrepreneurial, and institutional practices.

Singh (1994) highlights extension's focus on facilitating the transfer of technology to enhance standards of living.

Scope of Extension Education

Extension education encompasses a broad spectrum of areas, including agricultural production efficiency, marketing, natural resource management, farm and home management, family and youth development, leadership, community development, and public affairs. It serves as a mechanism for out-of-school education, empowering individuals to apply scientific knowledge to enhance their daily lives. Extension operates as a two-way channel, disseminating scientific information to rural communities while addressing their concerns through scientific solutions.

Career Opportunities in Extension Education

Career avenues in extension education span across government, private, and consultancy sectors, including roles such as Assistant Professors, Scientists, Subject Matter Specialists, and Consultants. Opportunities also exist in research projects, national and international NGOs, and self-employment ventures, offering diverse avenues for extension professionals to contribute to rural development and societal advancement.

In conclusion, extension education emerges as a dynamic field, rooted in disseminating knowledge and fostering development among rural communities, with vast opportunities for career growth and societal impact.

Extension Education Process

An effective extension educational program is founded upon five essential and interconnected steps, each integral to the success of the overall effort. While these steps are outlined separately for clarity, it is essential to recognize that they often occur concurrently and in tandem, with planning, teaching, and evaluation being ongoing processes throughout all phases of extension activities.

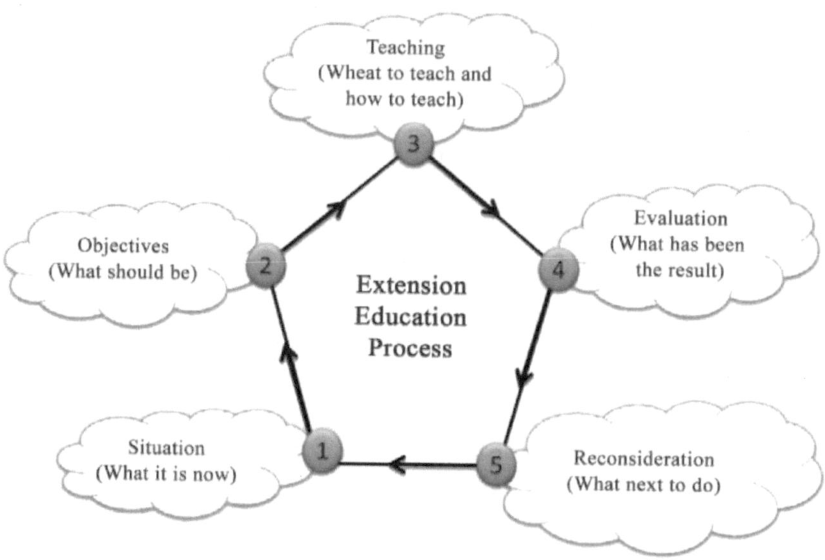

Extension Education is a participative & continuous process

First Step: Situation Analysis and Data Collection

The initial phase involves the collection of pertinent facts and the comprehensive analysis of the prevailing situation. This entails gathering information about the community's demographics, socio-economic status, cultural dynamics, physical environment, and technological landscape, alongside an in-depth understanding of their respective enterprises. These data are acquired through meticulous surveys and the establishment of strong rapport with the community members. Upon obtaining responses, a thorough analysis is conducted in collaboration with local stakeholders to identify prevalent issues and available resources within the community. For instance, following a comprehensive survey and subsequent data analysis, a prominent issue may be identified, such as the low income of farm families stemming from inadequacies in their crop production enterprises.

This initial step serves as the foundation upon which the subsequent phases of the extension educational program are built, providing crucial insights into the community's needs, challenges, and potential avenues for intervention. It underscores the importance of data-driven decision-making and community engagement in devising effective strategies for sustainable development and empowerment.

Second Step: Setting Realistic Objectives

Following the comprehensive analysis of the community's needs and resources, the next step involves the establishment of realistic objectives achievable by the community. This process entails the selection of a limited number of objectives in collaboration with local stakeholders. The objectives should be specific, clearly defined, and designed to bring satisfaction to the community upon completion. They should outline desired behavioral changes in individuals as well as the economic and social outcomes sought.

In the example provided, the identified problem of low income from crop production was further investigated, revealing factors such as low crop yields due to the use of local seeds with limited potential, minimal fertilizer application, and inadequate protection measures. Taking into account the community's capacity and available resources, the objective was set to increase crop yield by 20% within a specified timeframe. It was anticipated that the enhanced yield would lead to increased income, thereby improving family welfare.

Third Step: Implementation through Teaching

The third step involves the actual teaching process, encompassing the selection of teaching content, methods, and aids. This phase requires identifying research findings of economic and practical significance relevant to the community, as well as choosing appropriate teaching methods and aids.

In response to the identified problems, teaching content focused on technologies such as high-yielding variety (HYV) seeds, proper fertilizer application, and plant protection measures. Teaching methods included result and method demonstrations, farmers' training sessions, and dissemination through farm publications. Teaching aids such as tape recorders and slides were selected to facilitate learning.

Fourth Step: Evaluation of Teaching Effectiveness

Evaluation is essential to determine the extent to which the objectives have been achieved. Objective evaluation is facilitated through methods such as re-surveys, aiming to collect evidence of behavioral changes. This process not only measures success but also identifies any deficiencies that may exist.

In the example scenario, a re-survey conducted after the specified period revealed a 10% increase in crop yield, indicating a shortfall of 10% compared to the targeted 20%. Additionally, deficiencies were identified, including inadequate water management and financial constraints hindering the application of recommended inputs.

Fifth Step: Program Reconsideration and Adaptation

The final step involves re-evaluating the extension educational program in light of the evaluation results. Identified deficiencies become focal points for refining and adapting the program for continued progress. This iterative process ensures ongoing improvement and addresses emerging challenges or new developments.

Following reconsideration of the evaluation results with the community, revised teaching objectives were established. For instance, training sessions on proper water management practices and demonstrations were initiated. Additionally, the community was advised to seek timely production credit from banks to address input procurement challenges.

In essence, the cyclical nature of extension education perpetuates continuous improvement, guiding communities towards more favorable outcomes through informed decision-making and collaborative action.

3

Extension Education Objectives, Principles and Functions

Philosophy of Extension Education

Philosophy, derived from the Greek words "Philo" meaning love and "Sophia" meaning wisdom, is fundamentally the pursuit of wisdom and understanding. It encompasses the exploration of fundamental questions about existence, knowledge, values, reason, mind, and language. In a broader sense, philosophy serves as a comprehensive framework for seeking answers to life's most profound inquiries and understanding the nature of reality.

Within the context of specific disciplines, such as science, education, or ethics, philosophy plays a crucial role in providing foundational principles or guidelines. The philosophy of a particular discipline serves as a theoretical framework that informs and shapes the programs, activities, and practices within that field of knowledge. It elucidates the underlying assumptions, values, and methodologies inherent in the discipline, guiding practitioners in their pursuits and endeavors.

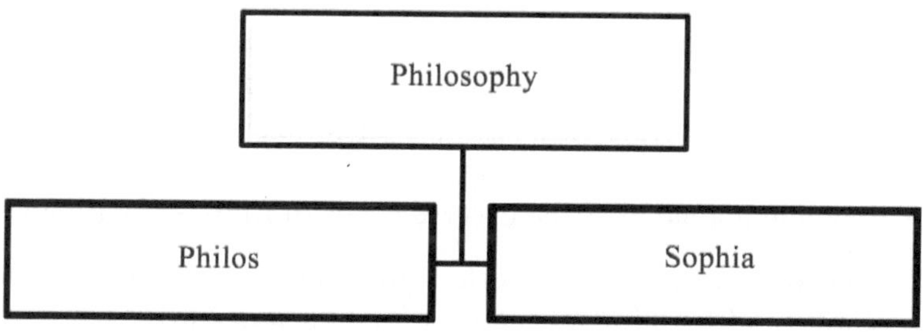

For example, the philosophy of education delineates the fundamental beliefs and principles underlying the process of teaching and learning. It addresses questions regarding the purpose of education, the nature of knowledge, the role of the teacher, and the desired outcomes of the educational process. By elucidating these foundational principles, the philosophy of education guides educators in designing curriculum, pedagogy, and assessment methods that align with their educational goals and values.

Similarly, in the realm of ethics, philosophical principles provide the framework for ethical decision-making and moral reasoning. Philosophical theories such as utilitarianism, deontology, and virtue ethics offer different perspectives on what constitutes ethical behavior and how moral dilemmas should be approached. By engaging with these philosophical frameworks, ethicists and practitioners in various fields can navigate complex ethical issues and formulate principled solutions.

In essence, the philosophy of a discipline serves as a compass, guiding practitioners and scholars in their pursuit of knowledge and understanding. By articulating fundamental principles and values, philosophy informs the theoretical underpinnings and practical applications within a given field, fostering coherence, rigor, and intellectual integrity.

The philosophy of extension work is rooted in the recognition of the pivotal role individuals play in fostering progress for rural communities and the nation at large. Extension educators adhere to the principle that empowering individuals is essential for societal advancement. They believe in working collaboratively with people to facilitate their development and enhance their overall well-being.

Central to the philosophy of extension work is the belief that rural communities possess inherent potential and resilience. Extension educators view themselves as facilitators, working alongside community members to identify their needs, aspirations, and challenges. By fostering partnerships and participatory approaches, they aim to empower individuals to become active agents of change within their communities.

Furthermore, the philosophy of extension work emphasizes the importance of holistic development. Extension educators recognize that progress encompasses not only economic growth but also social, cultural, and environmental dimensions. They advocate for approaches that promote sustainable development and address the diverse needs of rural populations.

Ultimately, the philosophy of extension work is grounded in a commitment to human dignity, equity, and social justice. Extension educators strive to create inclusive spaces where all individuals have the opportunity to thrive and contribute to the collective welfare. Through education, capacity building, and empowerment, they seek to enable individuals to realize their full potential and achieve superior well-being, thereby fostering sustainable development and progress for rural communities and the nation as a whole.

Objectives of Extension Education

Indeed, objectives serve as the guiding aims toward which our efforts in extension education are directed. The more specific objectives of extension education are as follows:

1. *To Assist People in Problem Discovery and Analysis:* Extension educators aim to facilitate individuals and communities in recognizing and understanding their challenges. By employing participatory methods, they assist people in identifying their felt needs and analyzing the underlying issues they face. This process lays the groundwork for targeted interventions and solutions.
2. *To Foster Leadership Development and Community Organization:* Extension education endeavors to cultivate leadership skills among individuals within communities. Educators work collaboratively with community members to empower them to take ownership of their development initiatives. Through capacity building and organizational support, extension programs aim to mobilize communities to address their own challenges effectively.
3. *To Disseminate Research Information in an Accessible Manner:* A key objective of extension education is to bridge the gap between research findings and practical application. Extension educators strive to communicate relevant research information in a manner that is understandable and applicable to the target audience. By translating complex scientific knowledge into accessible formats, they empower individuals to make informed decisions and adopt evidence-based practices.
4. *To Facilitate Resource Mobilization and Utilization:* Extension programs seek to support individuals and communities in mobilizing and leveraging both internal and external resources. Educators assist in identifying available resources within the community and facilitate access to additional resources as needed. This objective aims to enhance community capacity and resilience by optimizing resource utilization for sustainable development.
5. *To Collect and Transmit Feedback Information for Continuous Improvement:* Extension education emphasizes the importance of ongoing learning and adaptation. Educators engage in systematic feedback collection to assess the effectiveness of extension interventions and identify areas for improvement. By fostering a culture of reflection

and learning, extension programs strive to continuously enhance their impact and relevance in addressing community needs.

In summary, the specific objectives of extension education encompass empowering communities to understand and address their challenges, fostering leadership and organizational capacity, facilitating knowledge dissemination and resource mobilization, and promoting continuous learning and improvement. These objectives serve as the foundation for extension programs aimed at empowering individuals and communities to achieve sustainable development and improved well-being.

Principles of Extension Education

The principles of extension education serve as foundational guidelines to ensure effective and meaningful engagement with rural communities. These principles are derived from a deep understanding of the dynamics and intricacies of community development, emphasizing the importance of tailoring interventions to local contexts and fostering active participation and empowerment. The key principles of extension education are as follows:

(i) Principle of Need and Interest: Extension programs must be responsive to the unique needs and interests of the community. Programs should be tailored to address specific challenges and aspirations, recognizing that these may vary widely across different regions and demographics.

(ii) Principle of Grassroots Level Organization: Extension initiatives should be initiated and driven by local community groups. By working closely with grassroots organizations, extension educators can ensure that programs are contextually relevant and effectively address local priorities.

(iii) Principle of Cultural Difference: Extension work should respect and integrate the cultural values and customs of the community. Understanding the cultural context is essential for building trust and rapport with community members and facilitating meaningful engagement.

4. Principle of Cooperation and People's Participation:

Extension is a collaborative endeavor that relies on the active involvement and cooperation of community members. Programs should empower individuals to take ownership of their development initiatives and work together towards common goals.

5. Principle of Cultural Change

Extension education should be sensitive to existing cultural norms and traditions while also promoting positive change where necessary. Respectful engagement with local customs and beliefs is essential for fostering acceptance and sustainability.

6. Principle of Learning by Doing

Hands-on learning experiences are central to effective extension education. By engaging in practical activities, individuals gain valuable skills and knowledge, enabling them to implement new practices with confidence.

7. Principle of Trained Specialists

Extension professionals should possess specialized knowledge and skills relevant to the needs of the community. Continuous training and professional development are essential for ensuring the quality and effectiveness of extension interventions.

8. Principle of Adaptability in the Use of Extension Teaching Methods:

Extension programs should utilize a variety of teaching methods tailored to the preferences and learning styles of the community. Flexibility and adaptability are key to engaging diverse audiences and maximizing learning outcomes.

9. Principle of Leadership

Local leadership plays a crucial role in driving change within communities. Extension programs should empower and support local leaders to mobilize resources and advocate for community development initiatives.

10. Principle of Whole Family

Extension efforts should adopt a holistic approach that considers the needs and interests of entire families. By addressing the diverse needs of households, programs can achieve more sustainable and impactful outcomes.

11. Principle of Evaluation

Continuous evaluation is essential for assessing the effectiveness and impact of extension programs. Evaluation should focus on measuring changes in knowledge, skills, attitudes, and behaviors, guiding program refinement and improvement.

12. Principle of Satisfaction

The ultimate measure of success in extension education is the satisfaction and well-being of community members. Programs should strive to deliver tangible benefits and solutions that address the needs and aspirations of stakeholders.

13. Principle of Indigenous Knowledge

Indigenous knowledge systems are valuable resources that should be recognized and integrated into extension programs. Respect for local knowledge and practices enhances the relevance and effectiveness of interventions, fostering sustainability and cultural preservation.

4

Extension Programme Planning

Extension Programme

The initial stride in fostering rural development systematically entails crafting programs that are grounded in the genuine needs of the populace. The formulation of such programs, necessitating meticulous planning aligned with the indigenous needs perceived by the local community and in consonance with the broader national imperatives, stands as a pivotal duty incumbent upon extension professionals across all echelons—national, state, district, block, and village.

Program planning constitutes a nuanced process involving the deliberation and determination of the trajectory and magnitude of extension education endeavors, aimed at effectuating social, economic, and technological transformations.

Program planning epitomizes a multifaceted endeavor, encompassing a series of sequential steps inclusive of problem identification, delineation of desired outcomes, resource assessment, program implementation, and evaluation. In essence, it encapsulates a methodical procedure wherein collaboration with stakeholders is integral to recognizing areas of deficiency or challenges and exploring potential remedies.

As posited by Kelsey and Hearne (1967), an extension program is characterized as a comprehensive statement delineating the prevailing situation, objectives, associated problems, and proposed solutions. While it maintains a relative permanence, continual refinement is imperative to ensure alignment with evolving needs.

Leagans (1961) elucidates that an "extension program" embodies a delineated set of objectives derived from meticulous situational analysis, aimed at attainment through concerted extension teaching endeavors.

Similarly, Lawrence (1962) delineates an "extension program" as the aggregate of all activities undertaken by county extension services, encompassing program planning, formulation of a written program statement, development of a plan of work, program execution, outcome assessment, and evaluation.

Program planning, thus, emerges as a deliberative decision-making process, entailing a critical examination of extant challenges and the exploration of optimal solutions through collaborative endeavors aimed at fostering community growth and development.

Understanding key terms in program planning is essential for effective implementation and evaluation. Here are the relevant terms:

Program: A program is a documented statement delineating proposed developmental activities, addressing the problems they aim to resolve, and outlining the necessary actions and resources required.

Planning: Planning involves the systematic examination of past and present data to forecast future trends, guiding the determination of goals to be achieved based on these projections.

Plan: A plan serves as a schedule of development work, outlining various activities to be undertaken within a specified timeframe. It addresses questions such as what, why, how, and when, as well as who will execute the work and where it will be carried out.

Aims: Aims are overarching and broad statements indicating the general direction of activities. For example, "Improving the economic condition of farmers" represents an aim.

Objectives: Objectives are specific expressions of desired outcomes toward which efforts are directed. For instance, "Increasing rice yield by 30 percent" serves as an objective.

Goal: A goal represents the anticipated progress in a specific direction within a defined period. For example, "To increase rice yield by 10 quintals per hectare in the current year" outlines a goal.

Problem: A problem refers to an unresolved situation, condition, or issue identified through study, necessitating a change.

Solution: A solution entails a proposed course of action aimed at transforming an unsatisfactory condition into a more desirable one.

Project: A project entails a specification of work or procedures required to accomplish a particular objective.

Plan of Work: A plan of work outlines activities in a manner conducive to efficient program execution, detailing what tasks will be performed, by whom, how, and when.

Calendar of Work: A calendar of work delineates activities to be undertaken in a specific chronological sequence, facilitating organized and timely execution of tasks.

Objectives of Programme

The general objective of an extension programme is to influence people to transform their life in better way. The assumption is that there is a need for change and make people aware are of this, if they are not and to develop their needs. Important objectives of having a programme planningas per Kelsey and Hearne (1966) are as follows:

1. To ensure careful consideration of what is to be done and why.
2. To furnish a guide against which to judge all new proposals.
3. To establish objectives toward which progress can be measured and evaluated.
4. To have a means of choosing the important (deep rooted) from incidental (less important) problems; and the permanent from the temporary changes.
5. To develop a common understanding about the means and endsbetween functionaries and organizations.
6. To ensure continuity during changes of personnel.
7. To help develop leadership.
8. To avoid wastage of time & money and promote efficiency.
9. To justify expenditure and to ensure flow of funds.
10. To have a statement in written form for public use.

Principles of Extension Programme Planning

Extension programs are designed with the explicit aim of enhancing rural livelihoods through individual, group, and community interventions. The planning of extension programs is underpinned by several key principles, which remain relevant regardless of the demographic or economic characteristics of the target audience:

1. *Analysis-Based Approach:* Extension programs should be formulated based on a comprehensive analysis of past experiences, current realities, and anticipated future needs. This entails gathering sufficient information about the local populace and their circumstances, followed by a rigorous

analysis and interpretation in collaboration with community stakeholders to ascertain future requirements.

2. *Clear and Purposeful Objectives:* Extension programs should articulate clear and meaningful objectives that address significant needs within the community. These objectives should be selected judiciously and explicitly stated to ensure alignment with the overarching goal of meeting the community's needs.
3. *Prioritization of Resources:* Given the finite nature of available resources, extension programs should prioritize initiatives based on resource availability and time constraints. This necessitates careful consideration and strategic allocation of resources to tackle the most pressing issues.
4. *Resource Availability and Utilization:* Extension programs should transparently outline the availability and utilization of resources, including funds, facilities, supplies, and personnel. Clarity regarding resource allocation is essential for ensuring the practicality and effectiveness of program implementation.
5. *Consensus Building:* Extension programs should garner consensus across various levels of governance, from the local to the national level. Alignment and coherence between programs at different administrative tiers are critical to avoid conflicts and ensure seamless implementation.
6. *Community Involvement:* Effective extension programs actively engage local communities throughout the planning and implementation phases. Local participation fosters ownership and sustainability of initiatives, thereby enhancing their impact and relevance.
7. *Collaboration with Relevant Institutions:* Extension programs necessitate collaboration with diverse institutions and organizations to leverage expertise and resources. The program should delineate the roles and contributions of relevant stakeholders in achieving program objectives.
8. *Comprehensive Plan of Work:* Extension programs should outline a detailed plan of work, either as a standalone document or integrated within the program framework. This plan should elucidate the execution strategy and delineate responsibilities to ensure efficient implementation.
9. *Evaluation and Adaptation:* Extension programs should incorporate mechanisms for ongoing evaluation and periodic reassessment. Regular monitoring of program outcomes facilitates evidence-based decision-making and allows for timely adjustments to optimize effectiveness.

10. *Equitable Benefit Distribution:* Extension programs should prioritize equitable distribution of benefits among community members to mitigate social disparities. Attention should be given to addressing the needs of marginalized groups to ensure inclusivity and social cohesion within the community.

Extension Programme Planning Process

The process of planning extension programs is a systematic and dynamic endeavor aimed at addressing the evolving needs and challenges faced by rural communities. It involves a series of interconnected steps that guide the development, implementation, and evaluation of initiatives to facilitate community development. Below are the essential steps involved in the extension programme planning process:

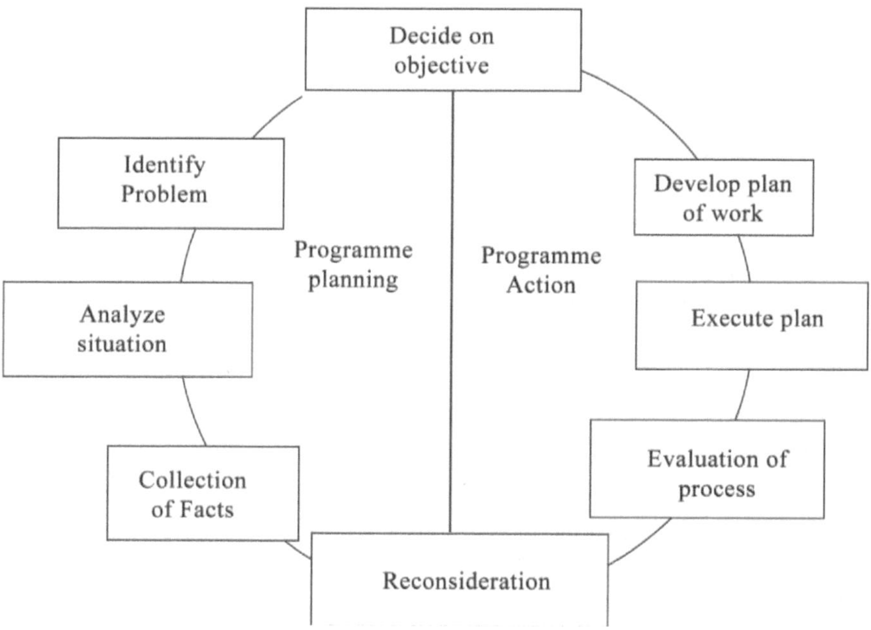

1. *Collection of Facts:* The foundation of effective planning lies in the comprehensive collection of relevant and reliable data. This entails gathering information about the demographic composition, socioeconomic status, prevailing agricultural practices, infrastructure, and institutional landscape of the target community. Data collection methods may include surveys, interviews, observations, and review of existing records and reports.

2. *Analysis of the Situation:* The collected data are meticulously analyzed and interpreted to discern patterns, trends, and underlying challenges within the community. This analytical process facilitates the identification of key issues and areas requiring intervention.
3. *Identification of Problems:* Through rigorous analysis, discrepancies between the current state ('what is') and the desired state ('what ought to be') are identified. These discrepancies represent the problems and needs of the community that necessitate attention and intervention.
4. *Determination of Objectives:* Based on the identified problems and needs, clear and specific objectives are formulated to guide the extension program. Objectives outline the desired outcomes and behavioral changes to be achieved within the community. They may encompass both short-term and long-term goals and must be articulated in measurable terms.
5. *Development of the Plan of Work:* An action plan is devised to operationalize the identified objectives and goals. This involves selecting appropriate strategies, activities, and interventions to address the identified problems. The action plan delineates responsibilities, timelines, and resource requirements for each activity, ensuring systematic implementation.
6. *Execution of the Plan of Work:* The action plan is set into motion through the implementation of planned activities and interventions. Efforts are made to mobilize necessary resources, including inputs, funding, and technical assistance, to support program implementation. Community participation is integral to this phase, fostering ownership and engagement.
7. *Evaluation:* Periodic evaluation is conducted to assess the effectiveness and impact of the extension program. Evaluation measures the extent to which program objectives have been achieved and identifies strengths, weaknesses, and areas for improvement. It encompasses both quantitative and qualitative assessments of program outcomes and processes.
8. *Reconsideration:* Based on evaluation findings, the extension program is reviewed and adjusted as necessary to enhance its efficacy and relevance. Reconsideration involves reflecting on the program's performance, identifying lessons learned, and making informed decisions for program refinement or adaptation. This iterative process ensures continuous improvement and responsiveness to evolving community needs.

In summary, extension programme planning is a cyclical and iterative process characterized by systematic data collection, analysis, objective setting, action planning, implementation, evaluation, and reconsideration. It embodies a commitment to participatory decision-making, community empowerment, and sustainable development.

5

Extension System in India Pre Independence Era

Rural development pertains to the enhancement of social, economic, and various other facets of rural life, primarily aimed at uplifting the rural poor. It encompasses a wide array of sectors and domains within rural communities. In its broadest sense, rural development entails progress across every aspect of rural existence. It is incumbent upon the government to undertake initiatives aimed at elevating the living standards of rural inhabitants. Over time, numerous rural development programs have been initiated by nationalists and social reformers. However, some of these programs have gradually phased out, while others have been assimilated into government-sponsored schemes. This attrition can be attributed to factors such as insufficient government support, financial constraints, and a dearth of adequately trained personnel. For the sake of clarity, rural development programs can be categorized into two distinct parts:

Stage-I	Pre-Independence Era (1866-1947)
Stage-II	Post-Independence Era (1947-1953)

Major Extension Programmes of Pre Independence Era (1866-1947)

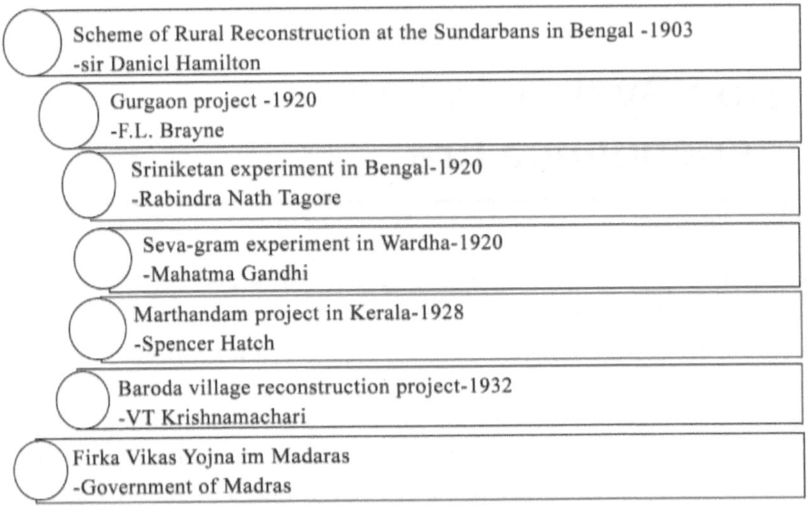

Year	Place	Person/agency
1903	Scheme of Rural Reconstruction at the Sundarbans in Bengal	Sir Daniel Hamilton
1920	Gurgaon project in Haryana	Mr. F.L. Brayne
1920	Sriniketan experiment in Bengal	Sri Rabindra Nath Tagore
1920	Seva-gram experiment in Wardha in Gujarat	Shree Mahatma Gandhi
1928	Marthandam project in Kerala	Dr. Spencer Hatch Young Men Christian Association (YMCA)
1932	Baroda village reconstruction project	Shree B.T. Krishnamachari
1942	Grow more food campaign	Agricultural departments of state governments
1945	Indian village service	Dr. W.H. Wisher
1946	Firka Vikas Yojana in Madras (now in Tamil Nadu)	Government of Madras

1. *Gurgaon Experiment:* Initiated in 1920 under the leadership of Mr. Brayne, who served as the Deputy Commissioner of Gurgaon district, the Gurgaon Project aimed to uplift rural communities comprehensively. This endeavor targeted advancements in agriculture, education, healthcare, sanitation, cooperation, and social welfare with fervor. Emphasizing the dignity of labor and self-reliance, the project utilized various communication channels such as films, songs, skits, and plays to promote increased agricultural yields and improved health standards. Although the project achieved some success, its continuity was disrupted upon Mr. Brayne's transfer.

2. *Sriniketan Project:* Founded by Rabindranath Tagore in 1920, the Sriniketan Institute for Rural Reconstruction, in collaboration with sociologist Shri L.M. Hurst, aimed to holistically enhance village life. The project focused on studying rural issues and aiding villagers in agricultural development, livestock improvement, and cooperative formation. Tagore advocated for self-help and community support, envisioning active participation of village workers in rural welfare activities. However, the project faced challenges such as the absence of market facilities, professional guidance, and inadequate incentives, hampering its success.

3. *Sewagram Experiment in Wardha:* The Sevagram project is a sustainable living model community in Maharashtra, India, created by Mahatma Gandhi. The project's goal is to make the extension system driven by farmers by providing new institutional arrangements for technology dissemination. The project aims to operationalize the extension reforms by establishing an Agricultural Technology Management Agency (ATMA) at the district level. The project also includes the KSHAMTA program, which involves KVKs across the country conducting 6009 on-farm trials and 34958 frontline demonstrations. The project also aims to provide a channel through which farmers' problems can be identified for research and for the modification of agricultural policies to the benefit of rural communities. The project also aims to give women a chance to participate in programs, as they have played an important role.

4. *Marthandam Project:* Established in 1921 by Dr. Spencer Hatch of the Young Men's Christian Association (YMCA), the Marthandam Project aimed at fostering comprehensive development—spiritual, mental, physical, economic, and social—among rural communities. Emphasizing self-help and self-reliance, the project aimed to eliminate poverty by promoting holistic growth. The Rural Demonstration Centre at Marthandam featured a demonstration farm, prized animals, and equipment for cottage industries, striving to uplift rural livelihoods.

5. *Baroda Village Reconstruction Project:* The Baroda experiment was a rural reconstruction program that began in 1932 and aimed to develop self-help and self-reliance. The Baroda rural reconstruction center was established in 1932 and began work in a group of villages around Kosamba. The center's goals included: Improving all aspects of rural life and Developing a "will to live better.

6. *Firka Development Scheme:* Introduced in Madras in 1946 as a government-sponsored initiative, the Firka Development Scheme aimed at organizing villagers for holistic development, both individually and as a community. It focused on areas with general backwardness, aiming to initiate cottage industries and improve living conditions. Short-term plans prioritized infrastructure and communication development, while long-term plans aimed at knowledge enhancement and the realization of Gandhian ideals.

Major Extension Programmes of Post-Independence Era (1947-1953)

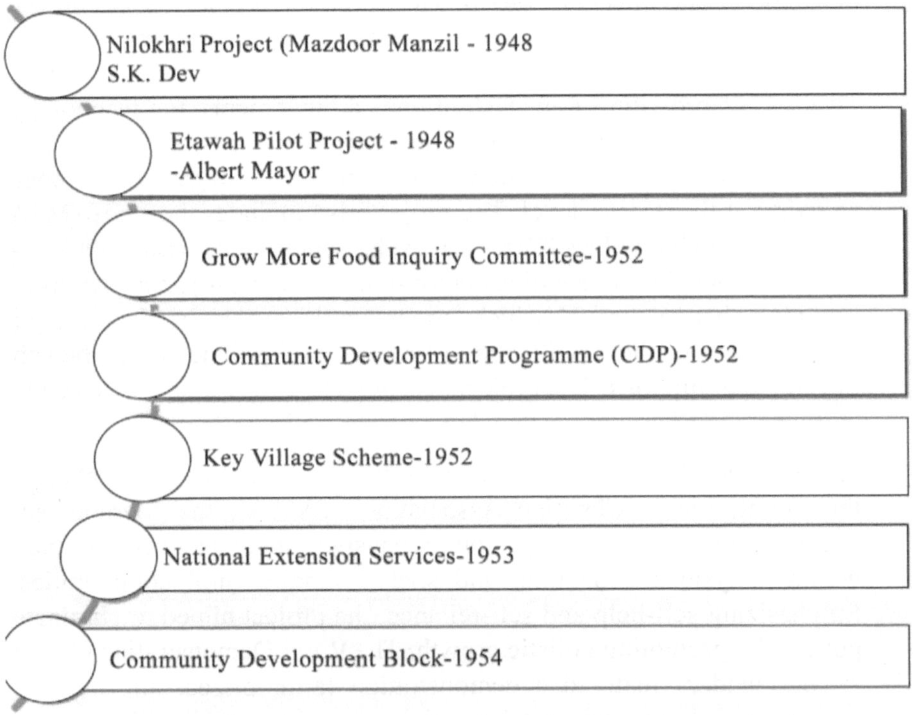

1. *Nilokheri Project:* Initiated by S.K. Dey, the Nilokheri Project aimed to rehabilitate approximately 7,000 displaced persons from Pakistan following the partition. Dey utilized 100 acres of swampy land located between Karnal and Kurukshetra for this purpose. The settlement, known as "Mazdoor Manzil," boasted various facilities including dairy, poultry, piggery, printing press, engineering workshop, and bone meal factory, all operated on cooperative principles. The project was founded on the principle of "he who would not work neither shall he eat."

2. *Etawah Pilot Project:* Following Independence in 1947, the Government of India prioritized rural development efforts and sought guidance from experienced individuals in this field. In response, the U.S. government dispatched Mr. Albert Mayor, a seasoned rural sociologist, to India. Under the leadership of Lt. Col. Albert Mayor, the Etawah Pilot Project office was established by the Uttar Pradesh government in October 1948, situated in Mahewa within the Etawah District. Initially, 64 villages surrounding Mahewa District were selected for development. Termed as the "Average District Plan," the project aimed to introduce development initiatives in a normal environment. Referred to as a pilot project, its purpose was to serve as a model for future community development initiatives. This program served as a precursor to the Community Development Programme (CDP).

Important Extension Programmes of Pre and Post Independence Era

S. No.	Name of the programme	Year	Started by	Place	Objectives
1.	Sriniketan	1920	Rabindra Nath Tagore	Sriniketan (Calcutta) West Bengal	• To help the rural people in establishing cottage industry. • To inspire the people to follow new technology. • Development of dairy farming. • To create the feeling of co-operation
2.	Gurgaon Project	1920	F.L Bryne	Gurgaon	• To increase crop production • To control extra expenditure • To improve the health. • To develop the feeling of women education
3.	Marthandam Project	1921	Dr. Spencer Hatch	Marthandam (Tamil Nadu)	• To develop the feeling of self reliance amongst the rural people. • To improve the spirit of cooperation and self-respect. • To bring the desirable change in the attitude of rural people.
4.	Firka Development	1946	Madras Government	Madras (Tamil Nadu)	• All-round development of rural people. • To develop the means of drinking water and communication. • To develop the committees of panchayat and co-operatives

S. No.	Name of the programme	Year	Started by	Place	Objectives
1.	Etawah Pilot Project	1948	Albert Mayor	Mahewa District- Etawah (U.P)	• To see what degree of production and social improvement can be obtain. • To find out the feeling of cooperation and initiative amongst the ruralpeople of an average area.
2.	Nilokheri Experiment	1948	S. K. Dey	Kurukshetra (Haryana)	• Self-dependence in all the fields of life. • To arrange for professional training and provide occupation for thepeople on the basis of their experience. • To eliminate the middle-man system

6

Extension Programmes Launched by ICAR / Govt. of India

India's agricultural landscape has witnessed a remarkable transformation, marked by a significant increase in food production owing to the Green Revolution. This achievement has been the result of a collaborative effort involving policy formulation, research endeavors, and extension support. Agricultural development in India is inherently location-specific, with the Union Government playing a guiding role in shaping policies to drive growth in the agricultural sector. While national-level programs are conceptualized by the central government, their implementation largely falls under the purview of various state government departments.

The Indian Council of Agricultural Research (ICAR) serves as the apex body overseeing research and extension activities aimed at fostering effective Transfer of Technology (TOT) models. Additionally, State Agricultural Universities contribute by developing extension models tailored to facilitate the transfer of technology, often leveraging frameworks established by the ICAR system.

Various agricultural development and extension programs, spearheaded by the Government of India and ICAR, have been instrumental in propelling the agricultural sector forward. These programs encompass a wide range of objectives and features, with some focusing on technological advancements (e.g., IADP, IAAP, ICDP, HYVP), while others prioritize social justice (e.g., MFAL, DPAP, T&V, IRDP, TRYSEM, NREP, DWACRA, NAEP, TMO, JRY, EAS, SFAC, SGSY).

In subsequent years, programs aimed at infrastructure development (e.g., NATP, PURA, NAIP, NREGA, NFSM) have gained prominence, alongside initiatives addressing changing climatic conditions and resource conservation (e.g., NICRA, FFS, SHC). Each new program has been designed to address gaps identified in previous initiatives, often leading to overlapping objectives. However, these programs collectively cater to various sections of the rural community, with a specific focus on the farming community.

Programs introduced in the 21st century incorporate ICT components to ensure the rapid dissemination of information. Noteworthy among these initiatives are:

Intensive Agricultural District Programme (IADP) - Launched in July 1960 across seven selected districts, later expanded to nine additional districts. The program aimed to increase productivity and production in high-potential districts, focusing on the adoption of improved agricultural practices.

Intensive Agricultural Area Programme (IAAP) - Implemented during 1964-65, IAAP emphasized the development of scientific and progressive agriculture in areas with high production potential. The program aimed to bring about 20 to 25 percent of the cultivated area under intensive agricultural development, with a focus on key crops like wheat, rice, millets, cotton, sugarcane, and pulses.

Difference between IADP and IAAP

IADP	IAAP
The main objective of the IADP was to implement the improved farm practices in the selected potential areas to increasethe production	The objective of the IAAP was to extend the concept of IADP to other potential areas to cover at least 20- 25% of the cultivated area in the country.
This programme was recommended by ateam of Food Ford foundation experts	This programme was recommended bythe third Five Year Plan mid-term appraisal committee.
The main criteria for selection of the districts was based on areas having assured irrigation facilities minimum natural hazards and well developed co- operatives	The main criteria for selection ofdistricts was based on areas having predominant crops and well developed infrastructure facilities
Twenty-eight districts including Mandyaand Raichur were covered under thisprogramme. Number of personnelworking in each community block were VLW's -10 AEO - 1 SMS - 2 (SMS were responsible for the wholedistrict)	One hundred and fifty districts including fourteen districts ofKarnataka were covered under theseprogrammes. Number of personnelworking in each community block were VLW's -10 AEO - 1 SMS – 2 (similar to IADP)

The All-India Coordinated Project on National Demonstration (1964) heralded a significant stride in agricultural productivity enhancement by leveraging proven agricultural technologies. Initiated by ICAR, this program aimed to demonstrate to farmers the potential of land productivity through the adoption of high-yielding crop varieties and comprehensive agricultural practices. Here are the key objectives and features of this program:

Objectives

1. To convincingly demonstrate to farmers the productivity potential of a unit area of land by employing high-yielding crop varieties and adopting comprehensive agricultural practices, including balanced fertilization and effective water management.
2. To showcase the use of implements for various operations and the utilization of soil testing laboratories for determining balanced fertilizer doses.
3. To utilize these demonstrations as effective training tools for farmers to encourage improved cultivation practices and serve as audio-visual aids for disseminating the latest research technology.
4. To provide researchers with firsthand knowledge of the challenges faced by farmers in adopting high-yielding varieties and identify constraints limiting crop production.
5. To minimize the time lag between research advancements and their practical application in the field.
6. To influence the extension systems of state agricultural departments and voluntary organizations by illustrating yield gaps and operational constraints.
7. To evaluate the income and employment generation potential of crops/ subjects under demonstration and educate farmers and extension agents accordingly.

While initial efforts focused on research station demonstrations, farmers expressed a preference for observing crop performance on their own fields. In response, the Ministry of Agriculture initiated the National Demonstration Project (NDP) in 1965-66. This nationwide initiative involved demonstrations conducted directly on farmers' fields, mainly by scientists from State Agricultural Universities (SAUs) and ICAR institutes. Key features of the National Demonstration Project include:

- Demonstrations conducted on farmers' fields, showcasing the potential of new seeds and agricultural practices.
- Area coverage ranging from 0.4 to 1.0 hectares per demonstration.
- Focus crops include wheat, paddy, sorghum, pearl millet, and maize.
- Demonstration duration of one year per crop, with a yield target of 9 tonnes of food grains per hectare.

- Financial support provided to farmers to cover the cost of critical inputs.
- Implementation at the district level by teams of subject matter specialists.
- Organizing field days to educate farmers about new agricultural technologies demonstrated on their fields.

In addition to the National Demonstration Project, other key agricultural development programs in India include the *High Yielding Variety Programme (HYVP-1966) and the Operational Research Project (1974-75).* These initiatives aim to enhance agricultural productivity, promote the adoption of modern technologies, and address socio-economic constraints hindering rapid agricultural advancement.

The High Yielding Variety Programme (HYVP-1966)

HYVP of 1966 played a pivotal role in India's journey towards achieving self-sufficiency in food production. Launched in 1966, this program was instrumental in ushering in the 'green revolution' by combining high-yielding crop varieties with modern agricultural practices. Here are the key highlights and objectives of HYVP-1966:

Key Points

1. The HYVP integrated high-yielding crop varieties with the application of balanced fertilizers, improved irrigation methods, plant protection measures, and enhanced agricultural implements. This holistic approach revolutionized agricultural productivity in the country.
2. The influence of high-yielding technology extended beyond cereal production to other sectors such as animal husbandry, fisheries, sericulture, and social forestry, fostering comprehensive agricultural development.
3. Initially launched in Punjab, Haryana, and the western parts of Uttar Pradesh, the program gradually expanded its reach to other regions, catalyzing widespread agricultural transformation.
4. Since its inception, the cultivation of high-yielding varieties has led to a significant increase in food grain production. Wheat production witnessed a doubling effect, while rice production also experienced substantial growth, albeit at a slightly lower rate.
5. The program surpassed its target of covering 2.5 crore hectares of land under high-yielding varieties of cereals and millets during the Fourth Five-Year Plan, ultimately achieving coverage exceeding four crore hectares.

Objectives

1. To achieve self-sufficiency in food production by leveraging high-yielding crop varieties along with modern agricultural inputs such as fertilizers, irrigation, plant protection measures, and improved implements.

Operational Research Project (1974-75)

In addition to HYVP-1966, the Operational Research Project (1974-75) aimed to further enhance agricultural productivity and address socio-economic constraints hindering rapid agricultural advancement. Here are the objectives of the Operational Research Project:

1. To test, adopt, and demonstrate new agricultural technologies directly on farmers' fields within entire villages or clusters of contiguous villages/ watershed areas.
2. To assess the profitability of these new technologies and evaluate their rate of adoption among farmers.
3. To identify both technological and socio-economic constraints that act as barriers to rapid agricultural change.
4. To demonstrate group action as a means of popularizing modern agricultural technologies at an accelerated pace, emphasizing collective participation and knowledge-sharing among farmers.

These initiatives underscored the commitment of the Indian government towards fostering agricultural innovation, enhancing productivity, and ensuring food security for the nation's growing population.

Krishi Vigyan Kendra (Farm Science Centre, 1974)

Krishi Vigyan Kendra (KVK) is a laudable initiative conceptualized and funded by the Indian Council of Agricultural Research (ICAR), New Delhi. These centers operate under the auspices of ICAR, State Agricultural Universities (SAUs), Central Universities, and Non-Governmental Organizations (NGOs). The core principle guiding KVKs is the dissemination of technology from laboratories to farmers' fields, known as the "lab to land" program. They focus on key areas of agriculture, including Crop Production, Plant Protection, Horticulture, Livestock Production and Management, Farm Engineering, Soil Water Conservation, Home Science, and allied fields. KVKs actively contribute to rural development by facilitating the transfer of technology.

Evolution of KVKs in India

The establishment of Krishi Vigyan Kendras was conceptualized based on recommendations from various governmental bodies and committees. The Education Commission (1964-66), Planning Commission, Inter-Ministerial Committee, and a committee led by Dr. Mohan Singh Mehta, appointed by ICAR in 1973, played significant roles in shaping this idea.

Objectives of KVKs

The overarching objectives of KVK activities are as follows

1. To demonstrate new and improved agricultural technologies directly to farmers and extension agencies in farmers' fields, fostering active participation.
2. To identify and prioritize key agricultural problems in the region based on farmers' needs and feedback from extension agencies.
3. To gather feedback from farmers and extension agencies, facilitating communication with research scientists for technology refinement.
4. To conduct training programs on various topics for different groups of villagers and extension functionaries.
5. To disseminate new and crucial information to extension agencies and NGOs for wider circulation in the locality, enhancing the economic conditions of the community.
6. To develop and validate extension models in farmers' fields with their participation, instilling confidence among them.

Activities undertaken by KVKs

KVKs engage in a range of activities in adopted villages to achieve their objectives, including:

a. Farm Advisory Service: Providing advisory services to farmers on various agricultural practices and technologies.

b. Training programs: Conducting training sessions for different categories of people, including farmers and extension workers.

c. Front Line Demonstrations (FLDs): Demonstrating new agricultural technologies directly in farmers' fields.

d. On Farm Testing (OFTs): Conducting trials of new agricultural practices and technologies on farmers' fields to assess their efficacy and suitability.

The Training and Visit (T & V) System in agricultural extension was pioneered by Daniel Benor and was first widely implemented in India by the World Bank in the early 1970s, following successful trials in Turkey. This system aimed to provide technical knowledge to farmers, with a focus on contact farmers who would then disseminate the information and guide others in adopting improved agricultural practices.

The T & V System operates through two main stages

1. *Training:* This stage involves the transfer of knowledge from subject matter specialists to extension workers.
2. *Visits:* After receiving training, extension workers visit farmers to transfer the knowledge acquired during training.

Objectives of the T & V System

1. Coordinate research, training, and extension activities effectively: The system aims to align research efforts with the needs and realities of farmers, ensuring that the information disseminated is relevant and useful.
2. Make research more effective: By focusing on local needs and conditions, the T & V System seeks to make research outcomes more impactful and applicable to farmers.
3. Evolve intensive training programs: The system aims to develop structured training programs for both extension workers and farmers, ensuring that they receive the necessary support and supervision to implement new practices effectively.

Organizational Structure of the T & V System

The organizational structure of the T & V System revolves around the number of farm families that a Village Extension Worker (VEW) can reasonably cover. This ensures that each extension worker has a manageable workload, allowing for effective communication and support to farmers.

Coverage of various extension personnel

Level	Extension personnel	Coverage
Field	Village Extension Worker (VEW)	800-1200 farm families
Block	Agricultural Extension Officer (AEO)	8 VEWs
Sub-Divisional	Sub-Divisional Extension Officer(SDEO)	5-8 AEOs
District	District Extension Officer (DEO)	4-8 SDEO in district

The Training and Visit (T & V) System in agricultural extension is characterized by several key features:

1. Professionalism: Extension agents undergo continuous training to fulfill their responsibilities in a professional manner, ensuring that they are equipped with the necessary knowledge and skills.
2. Single Line of Command: The extension service operates under a single line of technical and administrative command within the Ministry or Department of Agriculture, streamlining communication and decision-making processes.
3. Concentration of Effort: All extension staff focus solely on agricultural extension activities, ensuring that resources and efforts are directed towards improving agricultural practices.
4. Time-Bound Work: Messages and skills are communicated to farmers in a regular and timely manner. Village Extension Workers (VEWs) adhere to a fixed schedule, visiting farmers once every fortnight to provide support and guidance.
5. Field and Farmer Orientation: Contact with farmers occurs regularly and according to a known schedule, ensuring that a diverse range of farmers representing different farming and socio-economic backgrounds are reached.
6. Regular and Continuous Training: Fortnightly training sessions and monthly workshops are integral to the T & V System, providing opportunities to address farmer's problems, disseminate research findings, and develop production recommendations tailored to local conditions.
7. Linkages with Research: The T & V System maintains close linkages with agricultural research institutions through seasonal workshops, joint field visits, and training programs for extension staff. This ensures that research findings are effectively translated into practical recommendations for farmers.

Lab to Land Programme (LLP) (1979)

The Lab to Land Programme (LLP) and the National Agricultural Technology Project (NATP) were significant initiatives aimed at improving agricultural practices and the economic condition of farmers in India.

Objective: The main objective was to enhance the economic status of small and marginal farmers, as well as landless agricultural laborers, with a focus on scheduled castes and scheduled tribes, by transferring improved agricultural technologies.

Specific Objectives

1. Study and understand the background and resources of selected farmers and laborers.
2. Introduce low-cost and relevant agricultural and allied technologies to increase employment, production, and income.
3. Assist farmers in developing feasible farm plans tailored to their needs and resources.
4. Guide farmers in adopting improved technologies and demonstrate their economic viability.
5. Organize training programs and extension activities to prepare farmers for participation in agricultural development programs.
6. Make farmers aware of opportunities and agencies for economic advancement.
7. Develop linkages with scientists and institutions for guidance and advisory services and provide feedback to agricultural scientists and extension functionaries.

National Agricultural Technology Project (NATP) (1998)

Objective: NATP aimed to bring about significant changes in agricultural research and extension systems in India, with the support of the World Bank.

Components

1. Organization and Management System.
2. Agricultural Research.
3. Innovations in Technology Dissemination.

Project Implementing Agencies

1. Directorate of Extension (DOE).
2. National Institute of Agricultural Extension Management (MANAGE).
3. NATP cell at State Headquarters.
4. State Agricultural Management and Extension Training Institute (SAMETI).
5. District Level Agricultural Technology Management Agencies (ATMAs).

Both LLP and NATP were instrumental in introducing and disseminating improved agricultural technologies, providing training and extension services, and fostering collaborations between research institutions and farmers. These initiatives played a crucial role in enhancing agricultural productivity and improving the livelihoods of rural communities in India.

The NATP-ITD (Innovations in Technology Dissemination) implementation focused on piloting new approaches to technology transfer, organizational arrangements, and operational procedures in seven participating states in India. The purpose and goals of NATP-ITD were as follows:

Purpose

- Test new approaches to technology transfer, organizational arrangements, and operational procedures.

Goals

1. Decentralize decision-making to the district level by establishing Agricultural Technology Management Agency (ATMA) as a registered society.
2. Increase farmer input into program planning and resource allocation, especially at the block level, and enhance accountability to stakeholders.
3. Improve program coordination and integration.

The NATP-ITD scheme aimed to enhance the effectiveness of agricultural technology dissemination, improve coordination among various stakeholders, and foster partnerships to accelerate agricultural development at the grassroots level.

7

Development Programmes Launched by ICAR / Govt. of India

After the 1970s, the Indian Council of Agricultural Research (ICAR) and the Government of India launched several development programs to address various challenges in agriculture and rural development. These initiatives aimed to enhance agricultural productivity, improve rural livelihoods, ensure food security, and promote sustainable agriculture practices. Here's an introductory overview of some notable programs:

National Rural Employment Programme (1977)

The National Rural Employment Programme (NREP) of 1977 was a flagship initiative launched by the Indian government to alleviate poverty and unemployment in rural areas. It aimed to provide employment opportunities to the rural poor, particularly during the agricultural offseason when there was a lack of work.

Key features of the NREP included

1. *Employment Generation*: The program focused on creating wage employment for rural households through the implementation of various public works projects such as construction of roads, irrigation facilities, and rural infrastructure development.
2. *Target Beneficiaries*: The primary beneficiaries of the NREP were the marginalized sections of rural society, including landless laborers, small and marginal farmers, and other disadvantaged groups.
3. *Decentralized Implementation*: NREP emphasized decentralized planning and implementation, with significant decision-making power vested in local bodies like Gram Panchayats (village councils). This approach aimed to ensure that projects addressed the specific needs of local communities and provided employment opportunities where they were most needed.

4. *Focus on Asset Creation*: The program prioritized the creation of durable assets that would contribute to the long-term development of rural areas, such as rural roads, water conservation structures, and soil conservation measures.
5. *Subsidized Wages*: Workers engaged under NREP were provided with subsidized wages, ensuring that they earned a decent income while contributing to community development projects.

Overall, the National Rural Employment Programme of 1977 played a crucial role in addressing rural poverty and unemployment in India by providing sustainable employment opportunities and fostering rural development. It served as a precursor to subsequent rural employment schemes and contributed to the evolution of social welfare policies in the country.

Training of Rural Youth for Self-Employment (1979)

The Training of Rural Youth for Self Employment (TRYSEM) was launched in 1979 by the Government of India as a significant rural development initiative. The program aimed to provide skill training and entrepreneurship development opportunities to rural youth, particularly those from economically disadvantaged backgrounds, to enable them to generate self-employment opportunities.

Key features of TRYSEM included

1. *Skill Development*: The program focused on imparting practical skills and training in various trades and vocations relevant to rural livelihoods. This included training in agriculture, animal husbandry, handloom weaving, carpentry, masonry, and other traditional crafts.
2. *Entrepreneurship Development*: TRYSEM aimed to instill entrepreneurial spirit and capabilities among rural youth, equipping them with the knowledge and skills necessary to start and manage their own small businesses or enterprises.
3. *Target Beneficiaries*: The primary beneficiaries of TRYSEM were unemployed and underemployed rural youth, particularly school dropouts and those with limited formal education, who lacked access to meaningful employment opportunities.
4. *Financial Assistance*: The program provided financial assistance in the form of subsidized loans, grants, or other support mechanisms to enable trained youth to establish their own enterprises or engage in income-generating activities.

5. *Local Implementation*: TRYSEM emphasized decentralized implementation, with training centers established at the grassroots level in rural areas. This local focus ensured that training programs were tailored to the specific needs and requirements of each region.
6. *Integration with Rural Development*: TRYSEM was designed to complement other rural development initiatives, such as the National Rural Employment Programme (NREP), by providing a pathway for rural youth to transition from wage employment to self-employment and entrepreneurship.

Overall, the Training of Rural Youth for Self Employment (TRYSEM) program played a crucial role in promoting rural entrepreneurship and empowering rural youth by providing them with the necessary skills and resources to become self-reliant and economically independent.

Integrated Rural Development Programme (1980)

The Integrated Rural Development Programme (IRDP) of 1980 was a significant initiative launched by the Government of India to address rural poverty and promote holistic development in rural areas. It aimed to uplift the socio-economic conditions of the rural poor by integrating various development activities and providing them with sustainable livelihood opportunities.

Key features of the Integrated Rural Development Programme Included

1. *Target Beneficiaries*: The primary beneficiaries of IRDP were rural households living below the poverty line, particularly small and marginal farmers, landless laborers, and other disadvantaged groups.
2. *Integrated Approach*: IRDP adopted an integrated approach to rural development by combining various development interventions such as agriculture, animal husbandry, cottage industries, rural infrastructure, health, and education to address multiple dimensions of rural poverty.
3. *Subsidized Credit*: One of the key components of IRDP was the provision of subsidized credit to beneficiaries for undertaking income-generating activities and creating productive assets. This credit assistance helped rural households establish small businesses, purchase livestock, or invest in agriculture-related activities.
4. *Skill Training*: IRDP emphasized skill development and capacity building among rural beneficiaries to enhance their entrepreneurial capabilities and enable them to effectively manage income-generating activities. Training programs were conducted to impart technical and managerial skills relevant to rural livelihoods.

5. *Community Participation*: The program encouraged active participation of local communities, especially Gram Panchayats (village councils), in the planning, implementation, and monitoring of development activities. This participatory approach aimed to ensure that interventions were demand-driven and aligned with the needs and priorities of the rural population.
6. *Monitoring and Evaluation*: IRDP established mechanisms for regular monitoring and evaluation of project implementation to assess the impact of interventions, identify challenges, and make necessary course corrections for effective program delivery.

Overall, the Integrated Rural Development Programme (IRDP) of 1980 played a crucial role in promoting rural development, poverty alleviation, and empowerment of marginalized communities in India. While the program faced implementation challenges and criticisms over the years, it laid the foundation for subsequent rural development initiatives and contributed to the evolution of social welfare policies in the country.

Rural Landless Employment Guarantee Programme (1983)

The Rural Landless Employment Guarantee Programme (RLEGP) of 1983 was launched by the Government of India as a significant initiative to provide employment opportunities and alleviate poverty among landless rural households. The program aimed to guarantee a minimum number of days of employment to landless laborers in rural areas, thereby addressing their seasonal unemployment and underemployment.

Key features of the Rural Landless Employment Guarantee Programme included

1. *Target Beneficiaries*: RLEGP primarily targeted landless agricultural laborers and marginalized rural households who lacked access to productive assets and faced acute poverty due to seasonal unemployment.
2. *Employment Generation*: The program focused on creating wage employment opportunities for landless laborers through the implementation of various public works projects such as construction of rural infrastructure, soil conservation, water harvesting, and afforestation.
3. *Minimum Employment Guarantee*: RLEGP guaranteed a minimum number of days of employment to registered beneficiaries, typically ranging from 50 to 100 days in a year. This ensured a steady source of income for rural households during lean agricultural seasons when employment opportunities were limited.

4. *Wage Rates*: The program provided wages to workers at predetermined rates, which were usually higher than prevailing market wages, thereby improving the income levels and living standards of participating households.
5. *Community Participation*: RLEGP encouraged active participation of local communities, Gram Panchayats (village councils), and other stakeholders in the planning, implementation, and monitoring of public works projects. This participatory approach aimed to ensure transparency, accountability, and effective utilization of resources.
6. *Social Impact*: Apart from providing employment and income support, RLEGP contributed to the development of rural infrastructure, natural resource conservation, and environmental sustainability, thereby promoting overall socio-economic development in rural areas.
7. *Evaluation and Monitoring*: The program established mechanisms for monitoring and evaluation to assess the impact of employment generation activities, identify challenges, and make necessary improvements for effective program implementation.

Overall, the Rural Landless Employment Guarantee Programme (RLEGP) of 1983 played a crucial role in addressing rural poverty, seasonal unemployment, and underemployment among landless laborers in India. While the program faced implementation challenges and criticisms over the years, it contributed to the evolution of social welfare policies and rural development initiatives in the country.

Jawahar Rozgar Yojana (1989)

The Jawahar Rozgar Yojana (JRY) of 1989 was a flagship rural employment generation program launched by the Government of India. It aimed to provide sustainable livelihood opportunities and improve the socio-economic conditions of the rural poor by creating wage employment through public works projects and promoting self-employment ventures.

Key features of the Jawahar Rozgar Yojana included

1. *Convergence of Schemes*: JRY integrated various existing rural development programs, including the National Rural Employment Programme (NREP), Rural Landless Employment Guarantee Programme (RLEGP), and Self Employment Programme (SEP), to streamline implementation and maximize impact.

2. *Employment Generation*: The program focused on generating wage employment opportunities for rural households by implementing a range of public works projects such as construction of rural infrastructure (roads, bridges, culverts), soil conservation, water harvesting, and afforestation.
3. *Self-Employment Assistance*: JRY provided financial assistance, training, and technical support to promote self-employment ventures among rural youth, women, and other marginalized groups. This included support for activities such as small-scale industries, handicrafts, animal husbandry, and agriculture.
4. *Target Beneficiaries*: The primary beneficiaries of JRY were landless laborers, small and marginal farmers, women-headed households, and other vulnerable sections of rural society. The program aimed to prioritize the needs of the poorest households to ensure inclusive development.
5. *Community Participation*: JRY emphasized the participation of local communities, Gram Panchayats (village councils), and other stakeholders in the planning, implementation, and monitoring of development projects. This participatory approach aimed to empower rural communities and ensure the relevance and sustainability of interventions.
6. *Financial Support*: The program provided financial assistance to states and Union Territories for the implementation of rural development activities under JRY. Funds were allocated based on the assessed needs and priorities of each region, with a focus on backward and underdeveloped areas.
7. *Monitoring and Evaluation*: JRY established mechanisms for monitoring and evaluation to assess the progress, impact, and effectiveness of rural development initiatives. Regular reviews were conducted to identify challenges, address bottlenecks, and improve program delivery.

Overall, the Jawahar Rozgar Yojana (JRY) of 1989 played a significant role in promoting rural employment generation, poverty alleviation, and inclusive growth in India. While the program underwent several modifications and faced implementation challenges over the years, it contributed to the evolution of rural development strategies and social welfare policies in the country.

National Agriculture Extension Project (1982-83)

The National Agriculture Extension Project (NAEP) of 1982-83 was a significant initiative launched by the Government of India to strengthen agricultural extension services and improve technology dissemination to farmers across the country. It aimed to enhance agricultural productivity, promote modern farming practices, and increase the income levels of farmers through effective extension outreach.

Key features of the National Agriculture Extension Project included

1. *Enhanced Extension Services*: NAEP focused on revitalizing and modernizing agricultural extension services by strengthening the extension infrastructure, enhancing the capacity of extension personnel, and improving communication channels between agricultural scientists, extension workers, and farmers.
2. *Technology Transfer*: The project aimed to facilitate the transfer of agricultural technologies developed by research institutions to farmers' fields. This involved disseminating information on improved crop varieties, cultivation techniques, pest and disease management practices, soil conservation methods, and water management technologies through extension programs, demonstrations, and training sessions.
3. *Training and Capacity Building*: NAEP emphasized the training and capacity building of extension personnel, including agricultural scientists, extension officers, and village-level workers, to equip them with the knowledge and skills necessary to effectively communicate agricultural innovations and provide advisory services to farmers.
4. *Farmers' Participation*: The project encouraged active participation of farmers in the extension process by involving them in on-farm demonstrations, field trials, and farmers' field schools. This participatory approach aimed to promote farmer-to-farmer learning, adaptability of technologies to local conditions, and ownership of extension initiatives.
5. *Information Dissemination*: NAEP utilized various communication channels, including radio, television, print media, and mobile technology, to disseminate agricultural information, advisories, and best practices to a wide audience of farmers across different regions.
6. *Monitoring and Evaluation*: The project established mechanisms for monitoring and evaluating the effectiveness and impact of extension activities. Regular assessments were conducted to measure changes in agricultural productivity, adoption of new technologies, and improvement in farmers' incomes.

Overall, the National Agriculture Extension Project (NAEP) of 1982-83 played a crucial role in strengthening agricultural extension services, promoting technology adoption, and empowering farmers with knowledge and skills to enhance their livelihoods. It contributed to the modernization of agriculture and rural development in India, laying the foundation for subsequent extension programs and initiatives.

Watershed Development Programme (1994-95)

The Watershed Development Programme (WDP) of 1994-95 was launched by the Government of India with the aim of promoting sustainable land and water management practices in rainfed areas to enhance agricultural productivity, conserve natural resources, and improve the socio-economic conditions of rural communities.

Key features of the Watershed Development Programme included

1. *Integrated Watershed Management*: The program adopted an integrated approach to watershed management, focusing on the conservation and development of land, water, vegetation, and other natural resources within a defined geographical area.
2. *Community Participation*: WDP emphasized the active participation of local communities, including farmers, women, and other stakeholders, in the planning, implementation, and management of watershed development activities. This participatory approach aimed to ensure ownership, sustainability, and equitable distribution of benefits.
3. *Soil and Water Conservation*: The program implemented various soil and water conservation measures such as contour bunding, terracing, check dams, farm ponds, and afforestation to prevent soil erosion, recharge groundwater, and improve soil fertility and moisture retention.
4. *Livelihood Enhancement*: WDP integrated livelihood enhancement activities such as promotion of sustainable agriculture practices, agroforestry, livestock management, horticulture, and non-farm income generation activities to diversify rural livelihoods and improve household incomes.
5. *Capacity Building and Training*: The program provided training and capacity building support to farmers, watershed committees, and community-based organizations to enhance their knowledge and skills in watershed management, natural resource conservation, and sustainable agriculture practices.

6. *Convergence with Other Programs*: WDP promoted convergence with other rural development programs and initiatives such as Mahatma Gandhi National Rural Employment Guarantee Scheme (MGNREGS), Integrated Watershed Management Programme (IWMP), and National Rural Livelihoods Mission (NRLM) to leverage resources, expertise, and institutional support for effective implementation.

7. *Monitoring and Evaluation*: The program established mechanisms for monitoring and evaluation to assess the progress, impact, and sustainability of watershed development activities. Regular assessments were conducted to measure changes in land productivity, water availability, vegetation cover, and socio-economic indicators.

Overall, the Watershed Development Programme (WDP) of 1994-95 played a significant role in promoting sustainable land and water management practices, conserving natural resources, and improving the livelihoods of rural communities in India. It contributed to the resilience of rainfed agriculture, ecological restoration, and rural development efforts across the country.

Swarna Jayanti Gram Swarojgar Yojana (1999)

Swarn Jayanti Gram Swarojgar Yojana (SGSY) was launched in April 1999 by the Government of India. It aimed to alleviate poverty in rural areas by promoting self-employment and creating sustainable livelihood opportunities. The scheme primarily targeted the rural poor below the poverty line, with a focus on Scheduled Castes, Scheduled Tribes, and women.

Key Features of SGSY Included

1. Formation of Self Help Groups (SHGs): The scheme emphasized the formation of SHGs as the basic organizational unit. These groups facilitated access to credit, training, and technology for income-generating activities.

2. Skill Training: SGSY provided skill development and entrepreneurial training to enable beneficiaries to engage in various income-generating activities effectively.

3. Financial Assistance: Financial assistance was provided in the form of bank credit and government subsidies for setting up micro-enterprises and small businesses.

4. Infrastructure Development: The scheme also allocated funds for creating rural infrastructure like roads, irrigation facilities, and storage facilities to support economic activities.

5. Monitoring and Evaluation: SGSY implemented a robust monitoring and evaluation system to track the progress of beneficiaries and ensure effective utilization of resources.

Overall, SGSY aimed to empower rural communities by promoting self-employment, enhancing income levels, and reducing poverty through a holistic approach encompassing credit, training, and infrastructure development.

Agriculture Technology Management Agency (ATMA) – 2005

The Agriculture Technology Management Agency (ATMA) is a decentralized and farmer-oriented agricultural extension scheme launched by the Government of India. It was initiated in 2005 under the guidance of the Indian Council of Agricultural Research (ICAR) to promote sustainable agricultural development at the district level.

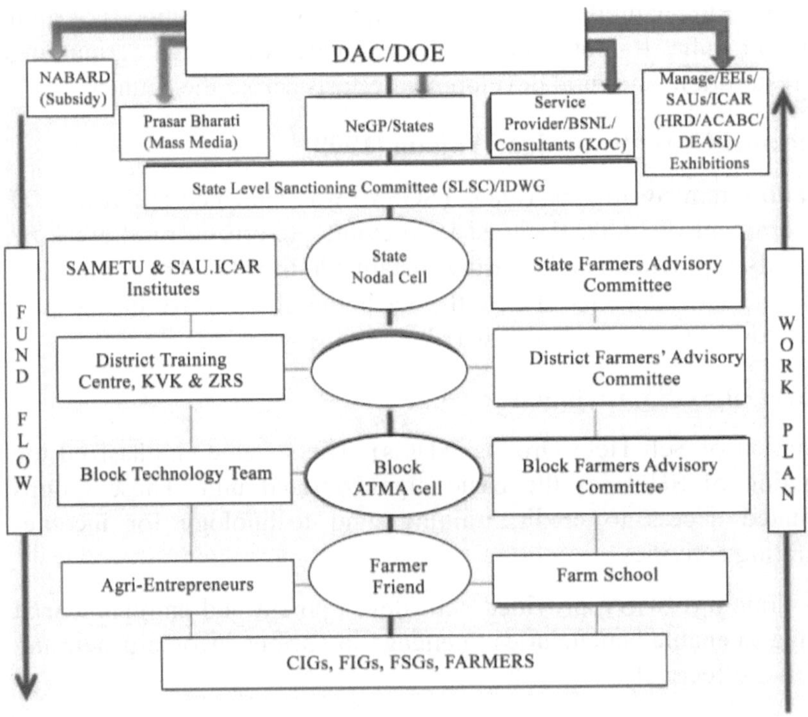

Organization Structure of ATMA

Key Features of ATMA

1. Decentralized Approach: ATMA operates at the district level, aiming to strengthen agricultural extension services by involving farmers, agricultural scientists, extension workers, NGOs, and other stakeholders.

2. Farmer-Centric: The agency focuses on addressing the needs and challenges faced by farmers, promoting farmer-centric agricultural development through the adoption of innovative technologies and best practices.
3. Capacity Building: ATMA conducts training programs, demonstrations, and workshops to enhance the skills and knowledge of farmers and extension personnel in modern agricultural practices, technology adoption, and entrepreneurship.
4. Technology Dissemination: It facilitates the dissemination of agricultural technologies, information, and market intelligence to farmers, enabling them to make informed decisions and improve their agricultural productivity and income.
5. Convergence and Coordination: ATMA promotes coordination and convergence among various agricultural development programs and stakeholders at the district level to ensure effective implementation and maximize impact.
6. Participatory Approach: The agency encourages the active participation of farmers in decision-making processes, planning, and implementation of agricultural development activities to ensure their ownership and sustainability.

The Strategic Research Extension Plan (SREP) in the Agriculture Technology Management Agency (ATMA) is a key component aimed at aligning research, extension, and farmer needs to promote sustainable agricultural development at the district level. SREP facilitates the integration of research findings into extension services and farmer practices to address local agricultural challenges effectively.

Here's how the Strategic Research Extension Plan (SREP) is Typically implemented within ATMA

1. Needs Assessment: The process begins with a comprehensive assessment of the agricultural needs, priorities, and challenges faced by farmers in the district. This assessment involves consultations with farmers, extension workers, agricultural scientists, and other stakeholders to identify key areas requiring research and technological interventions.
2. Identification of Researchable Issues: Based on the needs assessment, specific researchable issues and priority areas are identified. These may include crop productivity enhancement, pest and disease management,

soil health improvement, water management, post-harvest practices, and market access, among others.

3. Collaboration with Research Institutions: ATMA collaborates with agricultural research institutions, universities, and other relevant organizations to develop research proposals and projects tailored to address the identified issues. This collaboration ensures the utilization of scientific expertise, resources, and infrastructure for conducting applied research activities.
4. Technology Testing and Validation: Research activities under SREP involve field trials, demonstrations, and validation of technologies and practices suited to local agro-climatic conditions. These trials help assess the feasibility, efficacy, and adaptability of various interventions before their wider dissemination to farmers.
5. Farmer Participation and Feedback: Farmers are actively involved in the research process through on-farm demonstrations, participatory varietal selection, and farmer field schools. Their feedback and observations play a crucial role in evaluating the performance of technologies and refining research recommendations to suit local contexts.
6. Extension and Adoption: Once research findings are validated and proven effective, ATMA extension workers disseminate the technologies and practices to farmers through training programs, field days, farmer meetings, and extension materials. Extension efforts focus on building farmers' capacity, promoting adoption, and ensuring sustainable implementation of recommended practices.
7. Monitoring and Evaluation: SREP activities are monitored and evaluated regularly to assess the progress, impact, and effectiveness of research and extension interventions. Feedback from farmers, extension workers, and other stakeholders is incorporated into ongoing planning and implementation processes to improve outcomes.

By integrating research, extension, and farmer participation, the Strategic Research Extension Plan (SREP) in ATMA aims to foster innovation, enhance agricultural productivity, and improve farmers' livelihoods in the targeted districts.

Overall, ATMA plays a crucial role in bridging the gap between research and farmers, facilitating technology transfer, capacity building, and sustainable agricultural development at the grassroots level.

Mahatma Gandhi National Rural Employment Guarantee Act (MNREGA) 2006

The Mahatma Gandhi National Rural Employment Guarantee Act (MGNREGA) was enacted by the Government of India in 2005 and came into force in 2006. MGNREGA is one of the largest social welfare schemes in the world aimed at providing livelihood security to rural households in India.

Objectives

1. Guaranteeing Employment: The primary objective of MGNREGA is to provide at least 100 days of guaranteed wage employment in a financial year to every rural household whose adult members volunteer to do unskilled manual work.
2. Enhancing Livelihood Security: It aims to enhance the livelihood security of rural households by providing them with stable wage employment opportunities.
3. Strengthening Rural Infrastructure: MGNREGA focuses on creating durable assets in rural areas such as roads, water conservation structures, irrigation canals, and flood control measures, thereby contributing to rural development.

Key Provisions

1. Right to Employment: MGNREGA mandates the provision of legal entitlement to employment to rural households, ensuring that they have the right to demand work.
2. Unskilled Manual Work: The Act emphasizes the provision of unskilled manual work to ensure that a wide range of people, including the poor and marginalized sections of society, can participate in the program.
3. Wage Payment: Wages under MGNREGA are to be paid on a weekly basis and must be at least the minimum wage rate prescribed by the state government.
4. Participatory Planning: The Act emphasizes the role of local self-government institutions and Gram Sabhas in planning, implementation, and monitoring of projects undertaken under MGNREGA.
5. Social Audits: MGNREGA incorporates the provision for regular social audits to ensure transparency and accountability in the implementation of the program.

Impact

MGNREGA has had a significant impact on rural employment generation, poverty reduction, and empowerment of marginalized communities across India. It has helped in reducing distress migration from rural to urban areas by providing employment opportunities within villages. The scheme has also contributed to the creation of productive rural assets, thereby improving agricultural productivity and rural infrastructure.

Mahatma Gandhi National Rural Employment Guarantee Act (MGNREGA) stands as a landmark legislation in India's efforts towards poverty alleviation and rural development. By guaranteeing employment and creating rural assets, MGNREGA has played a crucial role in improving the lives and livelihoods of millions of rural households in India.

National Agricultural Innovation Project (2006)

The National Agricultural Innovation Project (NAIP) of 2006 was a major initiative launched by the Government of India with the aim of fostering innovation and enhancing agricultural productivity, sustainability, and profitability. The project sought to strengthen agricultural research, technology development, and knowledge dissemination systems to address the diverse challenges facing Indian agriculture.

Key Features of the National Agricultural Innovation Project Included

1. *Focus on Innovation*: NAIP focused on promoting innovation in agriculture by supporting research and development activities aimed at developing and adapting new technologies, practices, and solutions to improve crop yields, reduce production costs, and enhance resilience to climate change and other challenges.
2. *Partnership Approach*: The project emphasized collaboration and partnerships between various stakeholders, including agricultural research institutions, universities, private sector enterprises, farmers' organizations, and civil society organizations. These partnerships facilitated knowledge sharing, technology transfer, and capacity building across the agricultural innovation ecosystem.
3. *Research and Capacity Building*: NAIP supported research and capacity building initiatives in priority areas such as crop improvement, natural resource management, biotechnology, post-harvest management, value chain development, and agribusiness innovation. It aimed to strengthen

the scientific and technical capabilities of agricultural research institutions and human resources in the sector.

4. *Technology Transfer and Adoption*: The project focused on facilitating the adoption of innovative technologies and practices by farmers through extension services, demonstrations, training programs, and farmer field schools. It aimed to bridge the gap between research and practical application to ensure that farmers benefited from the latest agricultural innovations.

5. *Institutional Strengthening*: NAIP supported the strengthening of institutional mechanisms for agricultural research, extension, and innovation at the national, state, and local levels. It aimed to enhance the effectiveness, efficiency, and coordination of agricultural research and development efforts to better serve the needs of farmers and other stakeholders.

6. *Monitoring and Evaluation*: The project established mechanisms for monitoring and evaluating the progress, impact, and outcomes of agricultural innovation activities. Regular assessments were conducted to assess the adoption and impact of new technologies, as well as to identify lessons learned and best practices for future interventions.

Overall, the National Agricultural Innovation Project (NAIP) of 2006 played a crucial role in promoting agricultural innovation, technology transfer, and sustainable development in India. It contributed to strengthening the country's agricultural research and development ecosystem, fostering partnerships, and empowering farmers with knowledge and tools to improve their livelihoods and resilience in the face of evolving agricultural challenges.

National Innovations on Climate Resilient Agriculture (NICRA) – 2011

The National Innovations on Climate Resilient Agriculture (NICRA) was launched in 2011 by the Indian Council of Agricultural Research (ICAR) as a flagship initiative to address the challenges posed by climate change on agriculture. NICRA aimed to develop and promote climate-resilient agricultural practices and technologies to enhance the adaptive capacity of farmers and ensure sustainable agricultural production in the face of changing climatic conditions.

Key features of the National Innovations on Climate Resilient Agriculture included

1. *Climate Change Adaptation*: NICRA focused on developing and promoting climate-resilient agricultural practices, technologies, and cropping systems that are better suited to withstand the impacts of climate change, such as erratic rainfall, temperature fluctuations, droughts, floods, and pest and disease outbreaks.
2. *Research and Technology Development*: The project supported research activities aimed at understanding the climate change impacts on agriculture, identifying vulnerable regions and crops, and developing innovative solutions to mitigate risks and enhance resilience. This included crop breeding for drought tolerance, water-saving irrigation techniques, soil health management, agroforestry, and integrated pest management.
3. *On-Farm Demonstrations*: NICRA conducted on-farm demonstrations and pilot projects to showcase climate-resilient agricultural practices and technologies to farmers. These demonstrations served as learning platforms for farmers to adopt and replicate these practices on their own farms.
4. *Capacity Building*: The project provided training, capacity building, and extension support to farmers, extension workers, and other stakeholders to enhance their knowledge and skills in climate-resilient agriculture. Training programs covered topics such as climate-smart farming practices, weather forecasting, crop management, and risk mitigation strategies.
5. *Knowledge Sharing and Networking*: NICRA facilitated knowledge sharing, exchange of best practices, and collaboration among farmers, scientists, policymakers, and other stakeholders through workshops, seminars, field days, and farmer-to-farmer learning networks. This networking helped in scaling up successful climate-resilient agriculture interventions across different regions.
6. *Monitoring and Evaluation*: The project established monitoring and evaluation mechanisms to assess the adoption, impact, and effectiveness of climate-resilient agricultural practices and technologies. Regular assessments were conducted to measure changes in crop yields, income levels, and resilience indicators among participating farmers.

Overall, the National Innovations on Climate Resilient Agriculture (NICRA) of 2011 played a significant role in promoting climate-smart agriculture, enhancing farmers' adaptive capacity, and building resilience to climate change in India. It contributed to sustainable agricultural development, food security, and livelihood improvement in climate-vulnerable regions across the country.

Attracting and Retaining Youth in Agriculture (2015)

"Attracting and Retaining Youth in Agriculture" was an initiative launched in 2015 by various stakeholders including governments, international organizations, and NGOs, aimed at addressing the challenge of youth migration from rural areas and promoting youth engagement in agriculture. The initiative recognized the importance of involving young people in agriculture to ensure food security, promote rural development, and sustain agricultural productivity.

Key Geatures of the Initiative Included

1. *Youth Empowerment*: The initiative focused on empowering youth by providing them with access to education, training, and resources needed to engage in agriculture as a viable and attractive livelihood option. This included support for entrepreneurship, vocational training, and skill development tailored to the needs of rural youth.

2. *Technology and Innovation*: The initiative promoted the adoption of modern agricultural technologies, practices, and innovations to make farming more productive, efficient, and profitable. It encouraged the use of ICT tools, precision agriculture, mechanization, and value-added processing to attract tech-savvy youth to agriculture.

3. *Access to Finance and Land*: Recognizing the challenges young farmers face in accessing finance and land, the initiative advocated for policies and programs that facilitate access to credit, land tenure security, and agricultural inputs for young entrepreneurs. It also promoted innovative financing mechanisms such as youth-targeted loans and leasing arrangements.

4. *Market Linkages and Value Chains*: The initiative emphasized the importance of strengthening market linkages and value chains to enable young farmers to access markets, receive fair prices for their produce, and add value to agricultural products through processing and value addition. It promoted entrepreneurship and agribusiness development among youth.

5. *Policy Support and Advocacy*: The initiative advocated for supportive policies, incentives, and investments by governments and other stakeholders to create an enabling environment for youth engagement in agriculture. This included policy reforms, youth-targeted programs, and investment in rural infrastructure and services.

6. *Youth Participation and Networking*: The initiative encouraged youth participation in decision-making processes, policy dialogue, and agricultural organizations. It facilitated networking and knowledge exchange among young farmers, agripreneurs, and rural youth organizations to foster collaboration and peer learning.

Overall, the "Attracting and Retaining Youth in Agriculture" initiative aimed to harness the potential of young people as agents of change in agriculture and rural development. By addressing the specific needs and aspirations of rural youth, it sought to transform agriculture into a dynamic, inclusive, and sustainable sector that attracts and retains the next generation of farmers and agripreneurs.

Farmers FIRST Programme (Farm, Innovations, Resources, Science and Technology) – 2016

The Farmers FIRST Programme (Farm, Innovations, Resources, Science and Technology) was launched in 2016 by the Indian Council of Agricultural Research (ICAR) with the objective of addressing the key challenges faced by farmers and enhancing agricultural productivity, sustainability, and profitability through the adoption of innovative technologies and practices.

Key Features of the Farmers FIRST Programme Included

1. *Farmers-Centric Approach*: The program adopted a farmers-centric approach, focusing on understanding the needs, priorities, and constraints faced by farmers in different agro-climatic regions. It aimed to develop and disseminate location-specific technologies and solutions tailored to the requirements of farmers.

2. *Innovation and Technology Adoption*: Farmers FIRST emphasized the promotion and adoption of innovative technologies, practices, and approaches to address critical issues such as soil health management, water conservation, crop diversification, pest and disease management, and post-harvest management. It encouraged the use of science and technology to improve farm efficiency and sustainability.

3. *Participatory Research and Extension*: The program promoted participatory research and extension methodologies, involving farmers as partners in the co-creation and validation of agricultural innovations. It facilitated farmer-led demonstrations, field trials, and technology dissemination activities to promote the adoption of best practices.
4. *Capacity Building and Training*: Farmers FIRST focused on building the capacity of farmers, extension workers, and other stakeholders through training, workshops, exposure visits, and skill development programs. It aimed to enhance the knowledge and skills of farmers in modern agricultural practices, value addition, market linkages, and entrepreneurship.
5. *Resource Optimization*: The program emphasized the optimal utilization of natural resources such as land, water, and energy to enhance agricultural productivity while ensuring environmental sustainability. It promoted the adoption of resource-efficient technologies and conservation practices to minimize input costs and environmental impacts.
6. *Scaling-Up Successful Interventions*: Farmers FIRST facilitated the scaling-up of successful interventions and best practices through the replication and expansion of proven technologies and approaches across different regions. It aimed to maximize the impact of agricultural innovations and ensure widespread adoption among farmers.
7. *Monitoring and Evaluation*: The program established mechanisms for monitoring and evaluating the progress, impact, and outcomes of agricultural interventions. Regular assessments were conducted to measure changes in farm productivity, income levels, adoption rates, and sustainability indicators among participating farmers.

Overall, the Farmers FIRST Programme (Farm, Innovations, Resources, Science and Technology) of 2016 played a crucial role in promoting farmer-centric approaches to agricultural development, fostering innovation and technology adoption, and improving the livelihoods of farmers in India. It contributed to the transformation of Indian agriculture into a modern, sustainable, and resilient sector that meets the evolving needs of farmers and society.

Pradhan Mantri Fasal Bima Yojana (2016)

The Pradhan Mantri Fasal Bima Yojana (PMFBY) is a crop insurance scheme launched by the Government of India in 2016 with the objective of providing financial protection to farmers against crop loss due to natural calamities, pests, and diseases. The scheme aims to mitigate the financial risks faced by farmers and promote agricultural sustainability and resilience.

Key Features of the Pradhan Mantri Fasal Bima Yojana Include

1. *Comprehensive Coverage*: PMFBY provides comprehensive insurance coverage to farmers for all food crops, oilseeds, and horticultural crops grown in notified areas. It covers pre-sowing risks, post-harvest losses, and yield losses due to adverse weather events, pest attacks, diseases, and other specified perils.
2. *Premium Subsidy*: The scheme offers premium subsidies to farmers to make crop insurance affordable and accessible. The premium rates are heavily subsidized by the central and state governments, with farmers required to pay only a nominal premium amount based on the insured crop and coverage level.
3. *Risk Assessment and Mapping*: PMFBY utilizes advanced technologies such as remote sensing, satellite imagery, and crop cutting experiments (CCEs) to assess crop yield potential, monitor crop health, and estimate yield losses accurately. This helps in determining insurance payouts and compensating farmers for actual losses incurred.
4. *Timely Settlement of Claims*: The scheme emphasizes timely settlement of insurance claims to ensure prompt compensation to affected farmers. Claims are processed and settled within a specified timeframe, usually within a few weeks of crop loss assessment, to provide financial relief to farmers in distress.
5. *Digitization and Transparency*: PMFBY leverages digital platforms and online portals for enrollment, premium payment, insurance coverage validation, and claims processing. It promotes transparency, accountability, and efficiency in the administration of crop insurance schemes and minimizes paperwork and bureaucratic delays.
6. *Crop Diversification and Sustainable Agriculture*: The scheme encourages crop diversification and adoption of sustainable agricultural practices by providing insurance coverage for a wide range of crops. It incentivizes farmers to shift from water-intensive or risk-prone crops to resilient and economically viable cropping patterns.
7. *Awareness and Outreach*: PMFBY conducts awareness campaigns, training workshops, and outreach activities to educate farmers about the benefits of crop insurance, coverage options, claims procedures, and risk

mitigation strategies. It aims to increase the uptake of crop insurance among farmers and enhance their risk management capabilities.

Overall, the Pradhan Mantri Fasal Bima Yojana (PMFBY) plays a crucial role in protecting farmers' income and promoting agricultural resilience in the face of climate variability and natural disasters. By providing financial security and risk mitigation tools, the scheme contributes to the well-being and livelihoods of millions of farmers across India.

National Agricultural Higher Education Project (NAHEP)

The National Agricultural Higher Education Project (NAHEP) of 2017 is an initiative launched by the Government of India with the aim of enhancing the quality of agricultural higher education in the country. The project focuses on strengthening agricultural universities and institutions to produce skilled manpower, conduct cutting-edge research, and address the emerging challenges facing the agricultural sector.

Key features of the National Agricultural Higher Education Project Include

1. *Infrastructure Development*: NAHEP supports the development of infrastructure and facilities in agricultural universities and institutions to improve teaching, research, and extension activities. This includes the establishment of state-of-the-art laboratories, classrooms, libraries, and research farms.

2. *Faculty Development*: The project emphasizes the professional development of faculty members through training, workshops, seminars, and exchange programs. It aims to enhance their teaching and research capabilities, update their knowledge and skills, and promote interdisciplinary collaboration.

3. *Curriculum Reform*: NAHEP facilitates the revision and modernization of agricultural curricula to make them more relevant, interdisciplinary, and responsive to the changing needs of the agricultural sector. It introduces new courses, modules, and practical training components to align education with industry requirements.

4. *Technology Adoption*: The project promotes the adoption of modern teaching methods, pedagogical tools, and information and communication technologies (ICTs) in agricultural education. It encourages the use of e-learning platforms, virtual classrooms, and digital resources to enhance learning outcomes and accessibility.

5. *Research and Innovation*: NAHEP supports research and innovation activities in agricultural universities to address priority areas such as crop improvement, soil health management, water conservation, climate resilience, and value addition. It fosters collaboration with industry, government agencies, and international institutions to promote technology transfer and commercialization of research outcomes.
6. *Entrepreneurship Development*: The project encourages entrepreneurship and innovation among students and faculty members by providing incubation support, startup grants, and mentorship programs. It aims to nurture a culture of entrepreneurship and innovation in agricultural education institutions and facilitate the establishment of agribusiness ventures.
7. *Industry-Academia Linkages*: NAHEP facilitates closer collaboration between agricultural universities and industry stakeholders to bridge the gap between academia and the agriculture sector. It promotes industry internships, experiential learning, and joint research projects to foster technology transfer, skill development, and employment opportunities for students.

Overall, the National Agricultural Higher Education Project (NAHEP) of 2017 aims to transform agricultural higher education in India by enhancing its relevance, quality, and competitiveness. By strengthening institutions, empowering faculty and students, and promoting innovation and entrepreneurship, the project contributes to building a skilled workforce and fostering sustainable agricultural development in the country.

Pradhan Mantri Annadata Aay Sanrakshan Abhiyan (2018)

The Pradhan Mantri Annadata Aay Sanrakshan Abhiyan (PM-AASHA) is a scheme launched by the Government of India with the aim of ensuring remunerative prices to farmers for their produce. It was introduced to address the challenges faced by farmers in selling their crops at fair prices and to mitigate agrarian distress.

Key Features of the PM-AASHA Scheme Include

1. *Price Support Mechanism*: PM-AASHA employs various price support mechanisms such as Minimum Support Price (MSP), procurement by government agencies, and direct purchase from farmers to ensure that farmers receive remunerative prices for their crops.

2. *Three Sub-Schemes*: PM-AASHA comprises three sub-schemes: Price Support Scheme (PSS), Price Deficiency Payment Scheme (PDPS), and Private Procurement & Stockist Scheme (PPSS). Each sub-scheme is designed to address different aspects of price support and procurement.
3. *Procurement of Staple Crops*: The scheme primarily focuses on procuring staple crops such as wheat, rice, and pulses, which are essential for food security and have a significant impact on the livelihoods of millions of farmers across India.
4. *Market Intervention*: PM-AASHA intervenes in agricultural markets to stabilize prices and ensure fair returns to farmers. It provides a safety net to farmers by ensuring that they do not incur losses due to price fluctuations or market distortions.
5. *Benefits for Farmers*: Through PM-AASHA, farmers are assured of receiving MSP for their produce, which provides them with a reliable source of income and reduces their dependence on market fluctuations. It also instills confidence among farmers and encourages them to invest in agricultural activities.
6. *Implementation and Monitoring*: The scheme is implemented by various agencies at the central and state levels, including the Food Corporation of India (FCI), National Agricultural Cooperative Marketing Federation of India Ltd. (NAFED), and State Procurement Agencies (SPAs). Monitoring mechanisms are in place to ensure effective implementation and transparent procurement processes.

Overall, the Pradhan Mantri Annadata Aay Sanrakshan Abhiyan (PM-AASHA) aims to support farmers and protect their incomes by ensuring fair prices for their crops. By addressing the challenges of price volatility and market uncertainties, the scheme contributes to the welfare and well-being of farmers and promotes agricultural sustainability.

Poshan Ma (2018)

"Poshan Ma" is a flagship initiative launched by the Government of India in 2018 as part of the broader "Poshan Abhiyan" or National Nutrition Mission. The program focuses on improving the nutritional status of pregnant and lactating mothers (hence the name "Poshan Ma," which translates to "Nutrition Mother" in English) and aims to address maternal malnutrition and related health issues.

Key Features of the Poshan Ma Program Include

1. *Antenatal Care*: Poshan Ma emphasizes the importance of antenatal care for pregnant women, including regular health check-ups, monitoring of maternal health indicators, and access to essential healthcare services.
2. *Nutritional Support*: The program provides nutritional supplements and dietary advice to pregnant and lactating women to ensure they receive adequate nutrition during crucial stages of pregnancy and breastfeeding.
3. *Health Education*: Poshan Ma conducts health education sessions and awareness campaigns to promote maternal nutrition, hygiene practices, breastfeeding, and the importance of immunization for both mothers and newborns.
4. *Community Participation*: The program encourages community participation and engagement through the involvement of frontline health workers, community health volunteers, and local self-help groups in delivering maternal healthcare services and disseminating nutrition-related messages.
5. *Monitoring and Evaluation*: Poshan Ma implements monitoring and evaluation mechanisms to track the progress and impact of maternal nutrition interventions. Regular assessments are conducted to measure changes in maternal health outcomes and nutritional status among program beneficiaries.

Overall, the Poshan Ma program aims to empower pregnant and lactating mothers with the knowledge, resources, and support they need to ensure a healthy pregnancy, safe childbirth, and optimal nutrition for themselves and their infants. By addressing maternal malnutrition and promoting maternal health and well-being, the program contributes to the broader goal of reducing maternal and child mortality and improving the overall nutritional status of women and children in India.

Soil Health Card Scheme (2019)

The Soil Health Card (SHC) Scheme is a government initiative launched in 2015 by the Government of India with the aim of assessing soil health and providing personalized recommendations to farmers for improving soil fertility and productivity. The scheme seeks to promote sustainable agriculture practices by empowering farmers with knowledge about the health of their soil and guiding them in making informed decisions regarding nutrient management and crop selection.

Key Features of the Soil Health Card Scheme Include

1. *Soil Testing*: Under the scheme, soil samples are collected from farmers' fields and analyzed in accredited laboratories to assess key soil parameters such as pH levels, nutrient content (nitrogen, phosphorus, potassium), organic carbon, and micronutrients. The results are used to generate Soil Health Cards for individual farmers.
2. *Personalized Recommendations*: Based on the soil test results, personalized recommendations are provided to farmers regarding appropriate doses and types of fertilizers, soil amendments, and crop-specific nutrient management practices. This helps farmers optimize nutrient use efficiency and reduce input costs while maximizing crop yields.
3. *Periodic Renewal*: Soil Health Cards are periodically renewed (usually every 2-3 years) to reflect changes in soil health due to farming practices and crop rotations. Farmers receive updated recommendations to ensure continuous improvement in soil fertility and sustainable agricultural production.
4. *Technology Adoption*: The scheme leverages information and communication technologies (ICTs) to streamline the soil testing process, manage soil health data, and disseminate soil health information to farmers. Mobile apps and online portals are used for soil sample registration, result dissemination, and farmer advisory services.
5. *Capacity Building*: The Soil Health Card Scheme includes capacity building initiatives such as training programs, workshops, and extension activities to raise awareness among farmers about the importance of soil health management, sustainable farming practices, and the use of Soil Health Cards as decision support tools.
6. *Incentives and Subsidies*: The government provides financial incentives and subsidies to farmers for soil testing and the adoption of recommended soil health management practices. This encourages farmers to actively participate in the scheme and invest in soil health improvement measures.

Overall, the Soil Health Card Scheme aims to empower farmers with knowledge and tools to make informed decisions about soil management and fertilizer use, thereby promoting sustainable agriculture, enhancing crop productivity, and safeguarding environmental health. By prioritizing soil health as a critical aspect of agricultural sustainability, the scheme contributes to the long-term resilience and viability of India's agricultural sector.

Pradhan Mantri Kisan Samman Nidhi Yojana (PM-KISAN) – 2019

The Pradhan Mantri Kisan Samman Nidhi (PM-KISAN) is a government scheme launched in 2019 by the Government of India with the aim of providing financial support to small and marginal farmers across the country. The scheme seeks to supplement the income of farmers and alleviate agrarian distress by providing direct income support through cash transfers.

Key features of the PM-KISAN scheme include

1. *Financial Assistance*: Under PM-KISAN, eligible farmers receive direct income support of Rs. 6,000 per year in three equal installments of Rs. 2,000 each. The financial assistance is provided directly into the bank accounts of beneficiary farmers through Direct Benefit Transfer (DBT) mode.
2. *Target Beneficiaries*: The scheme primarily targets small and marginal farmers who own cultivable land up to 2 hectares. Land-owning farmer families are identified and registered for the scheme based on land records maintained by state governments.
3. *Universal Coverage*: PM-KISAN aims to cover all eligible farmer households across the country, regardless of their socio-economic status or geographical location. The scheme is implemented pan-India, including all states and union territories, to ensure equitable distribution of benefits.
4. *Online Registration and Monitoring*: Farmers can register for PM-KISAN through an online portal or at designated Common Service Centers (CSCs) established at the village level. The scheme incorporates robust mechanisms for monitoring and verification of beneficiary data to prevent duplication and ensure transparency in implementation.
5. *Income Support for Agricultural Activities*: The financial assistance provided under PM-KISAN is intended to supplement the income of farmers and support them in meeting various expenses related to agricultural activities such as purchase of seeds, fertilizers, pesticides, and other inputs.
6. *Incentive for Sustainable Farming*: PM-KISAN encourages farmers to adopt sustainable farming practices and improve productivity by providing them with a stable source of income. The scheme aims to enhance the financial resilience of farmers and reduce their dependence on informal sources of credit.

Overall, the Pradhan Mantri Kisan Samman Nidhi (PM-KISAN) scheme plays a crucial role in providing direct income support to small and marginal farmers, promoting agricultural livelihoods, and ensuring inclusive growth in rural areas. By addressing the income needs of farmers, the scheme contributes to poverty reduction, food security, and rural development in India.

Pradhan Mantri Krishi Sinchai Yojana (2019)

The Pradhan Mantri Krishi Sinchai Yojana (PMKSY) is a flagship irrigation scheme launched by the Government of India in 2015, with an updated version introduced in 2019. It aims to improve water use efficiency, expand irrigation coverage, and enhance agricultural productivity across the country.

Key features of the Pradhan Mantri Krishi Sinchai Yojana (PMKSY) include

1. *Har Khet Ko Pani (Water to Every Field)*: The scheme focuses on ensuring the provision of water to every field (Har Khet Ko Pani) by promoting the development of efficient water management infrastructure such as water harvesting structures, irrigation systems, and micro-irrigation techniques.
2. *Multi-Pronged Approach*: PMKSY adopts a multi-pronged approach that includes the development of water resources, command area development, distribution infrastructure, and efficient water use practices. It aims to address the entire spectrum of irrigation needs, from water source augmentation to on-farm water management.
3. *Funding and Support*: The scheme provides financial assistance and support to state governments for the implementation of irrigation projects and water management initiatives. It encourages convergence with other government programs and funding sources to maximize resources and impact.
4. *Micro-Irrigation Promotion*: PMKSY promotes the adoption of micro-irrigation techniques such as drip irrigation and sprinkler irrigation to improve water use efficiency, reduce water wastage, and increase crop yields. Subsidies and incentives are provided to farmers for installing micro-irrigation systems.
5. *Watershed Development*: The scheme includes components for watershed development and rainwater harvesting to enhance water availability and recharge groundwater resources. It supports the implementation of soil and water conservation measures, afforestation, and integrated watershed management practices.

6. *Technology Adoption*: PMKSY encourages the adoption of modern technologies and practices for efficient water management and irrigation. It promotes the use of remote sensing, geographic information systems (GIS), and other digital tools for monitoring water resources and optimizing irrigation scheduling.
7. *Capacity Building and Awareness*: PMKSY includes provisions for capacity building, training, and awareness programs to educate farmers and stakeholders about water-saving techniques, sustainable irrigation practices, and the importance of water conservation.

Overall, the Pradhan Mantri Krishi Sinchai Yojana (PMKSY) of 2019 aims to transform India's agricultural landscape by ensuring equitable access to water resources, enhancing irrigation efficiency, and promoting sustainable water management practices. By improving water availability and productivity, the scheme contributes to food security, rural livelihoods, and agricultural growth in the country.

e-NAM (National Agricultural Market) 2019

The e-National Agricultural Market (e-NAM) is an online trading platform launched by the Government of India in 2016 with the aim of integrating agricultural markets across the country. It facilitates electronic trading of agricultural commodities and promotes transparent price discovery, competitive bidding, and efficient market operations.

Key features of the e-NAM platform include

1. *Unified Market Platform*: e-NAM provides a unified national market platform that connects multiple mandis (market yards) and Agricultural Produce Market Committees (APMCs) across different states and regions. It allows farmers to access a wider market reach and sell their produce to buyers from anywhere in the country.
2. *Transparent Price Discovery*: The platform offers real-time price information and market data to farmers, traders, and other stakeholders, enabling them to make informed decisions about buying and selling agricultural commodities. This promotes transparency in price discovery and reduces information asymmetry in agricultural markets.
3. *Online Trading*: e-NAM enables electronic trading of agricultural commodities through an online bidding process. Farmers can list their produce on the platform, set minimum reserve prices, and receive bids from registered buyers. The platform facilitates transparent, competitive, and hassle-free trading transactions.

4. *Quality Assurance*: e-NAM incorporates quality testing and certification mechanisms to ensure that agricultural commodities traded on the platform meet prescribed quality standards. This helps in building trust among buyers and ensures the quality and authenticity of traded produce.
5. *Reduced Intermediaries*: The platform aims to reduce the role of intermediaries in agricultural marketing by enabling direct interaction between farmers and buyers. It eliminates the need for middlemen and commission agents, thereby reducing transaction costs and increasing farmers' share in the market price.
6. *Ease of Access*: e-NAM is accessible through web portals and mobile applications, making it convenient for farmers, traders, commission agents, and other stakeholders to participate in online trading activities. It promotes inclusive participation and democratizes access to agricultural markets.
7. *Capacity Building*: The platform includes capacity building initiatives such as training programs, workshops, and awareness campaigns to familiarize farmers and other users with the e-NAM system, online trading procedures, and best practices in agricultural marketing.

Overall, the e-National Agricultural Market (e-NAM) plays a significant role in modernizing India's agricultural marketing infrastructure, promoting fair trade practices, and enhancing farmers' income by providing them with access to transparent, efficient, and inclusive agricultural markets.

National Bee Keeping and Honey Mission

The National Beekeeping and Honey Mission (NBHM) is a government initiative launched in 2020 by the Government of India. The mission aims to promote scientific beekeeping, enhance honey production, and improve the livelihoods of beekeepers across the country.

Key features of the National Beekeeping and Honey Mission (NBHM) include

1. *Promotion of Beekeeping*: NBHM aims to promote beekeeping as a sustainable and profitable livelihood option among farmers and rural communities. It provides training, technical support, and capacity building initiatives to encourage the adoption of modern beekeeping practices.

2. *Expansion of Honey Production*: The mission focuses on increasing honey production by leveraging the potential of beekeeping. It supports the establishment of new bee colonies, the adoption of high-yielding bee species, and the use of modern hive management techniques to enhance productivity.
3. *Quality Enhancement*: NBHM emphasizes the production of high-quality honey that meets national and international standards. It promotes good beekeeping practices, hygiene standards, and quality control measures to ensure the purity and safety of honey produced in India.
4. *Market Linkages and Value Addition*: The mission facilitates market linkages for beekeepers and honey producers, enabling them to access domestic and international markets. It promotes value addition through the processing, packaging, and branding of honey and bee-based products to increase their marketability and value.
5. *Pollination Services*: NBHM recognizes the importance of bees in pollination and ecosystem services. It promotes the conservation of bee habitats, biodiversity, and ecosystem resilience through sustainable beekeeping practices and habitat restoration initiatives.
6. *Financial Support and Incentives*: The mission provides financial assistance, subsidies, and incentives to beekeepers for the adoption of modern beekeeping technologies, purchase of equipment, and establishment of bee colonies. It also offers support for infrastructure development, honey processing units, and marketing facilities.
7. *Research and Development*: NBHM encourages research and development activities in beekeeping, honey production, and pollinator biology. It supports scientific studies, innovation, and technology transfer to address challenges and opportunities in the beekeeping sector.

Overall, the National Beekeeping and Honey Mission (NBHM) plays a crucial role in promoting sustainable agriculture, enhancing rural livelihoods, and conserving biodiversity. By harnessing the potential of beekeeping, the mission contributes to food security, environmental sustainability, and economic development in India.

Pradhan Mantri Garib Kalyan Package (PMGKP)

The Pradhan Mantri Garib Kalyan Package (PMGKP) is a relief package launched by the Government of India in response to the economic challenges posed by the COVID-19 pandemic. It was first introduced in March 2020 and has since undergone multiple iterations to provide support to vulnerable sections of society.

Key features of the Pradhan Mantri Garib Kalyan Package (PMGKP) include

1. *Food Security Measures*: PMGKP includes measures to ensure food security for vulnerable populations, such as the distribution of free food grains to beneficiaries of the Public Distribution System (PDS) and additional allocations of food grains to states and union territories.

2. *Cash Transfers*: The package provides direct cash transfers to various groups, including farmers, women, senior citizens, and differently-abled individuals, through the Direct Benefit Transfer (DBT) mechanism. This aims to provide immediate financial assistance to those in need.

3. *Healthcare Support*: PMGKP includes provisions for strengthening the healthcare infrastructure and providing support to frontline healthcare workers. It allocates funds for the procurement of medical equipment, testing kits, and personal protective equipment (PPE) to combat the COVID-19 pandemic effectively.

4. *Insurance Coverage*: The package offers insurance coverage to healthcare workers, including doctors, nurses, and sanitation workers, who are at the forefront of the fight against COVID-19. This includes compensation in case of accidental loss of life due to COVID-19-related duties.

5. *Support for Women and Children*: PMGKP provides additional support to women and children through various schemes, including the provision of free cooking gas cylinders, increased allocations for the Women and Child Development Ministry's programs, and support for anganwadi workers.

6. *Employment Generation*: The package includes measures to boost employment generation and livelihood opportunities for migrant workers and other vulnerable groups. This includes increased allocations for the Mahatma Gandhi National Rural Employment Guarantee Act (MGNREGA) and other rural development programs.

Overall, the Pradhan Mantri Garib Kalyan Package (PMGKP) aims to provide immediate relief to the most vulnerable sections of society affected by the COVID-19 pandemic. By addressing food security, healthcare needs, financial assistance, and employment generation, the package seeks to mitigate the socio-economic impact of the crisis and ensure the well-being of the marginalized and disadvantaged populations in India.

Knowledge Systems and Homestead Agriculture Management in Tribal Areas 2022

"Knowledge Systems and Homestead Agriculture Management in Tribal Areas" refers to an integrated approach aimed at promoting sustainable agricultural practices and preserving traditional knowledge systems among tribal communities. This initiative recognizes the unique cultural and ecological context of tribal areas and seeks to enhance food security, livelihoods, and resilience among tribal households through the following key components:

1. *Preservation of Traditional Knowledge*: Recognizing the rich indigenous knowledge systems present within tribal communities regarding agriculture, biodiversity conservation, and natural resource management. Efforts are made to document, preserve, and promote these traditional practices, which have been developed and refined over generations.
2. *Capacity Building and Training*: Providing training, workshops, and capacity-building programs to tribal farmers on sustainable agricultural practices tailored to their local conditions. This includes techniques such as agroforestry, organic farming, crop diversification, and soil and water conservation methods.
3. *Homestead Agriculture Promotion*: Encouraging the establishment of homestead gardens and small-scale agricultural enterprises within tribal households. This involves cultivating a variety of crops, vegetables, fruits, and medicinal plants in and around the homestead to meet nutritional needs and generate supplementary income.
4. *Ecosystem Conservation*: Emphasizing the importance of preserving local ecosystems, biodiversity, and natural resources in tribal areas. Efforts are made to promote sustainable land use practices, forest conservation, watershed management, and the protection of traditional agroecological landscapes.
5. *Livelihood Enhancement*: Supporting tribal communities in accessing markets, credit, and extension services to improve their income from agricultural produce and non-timber forest products (NTFPs). This may involve facilitating market linkages, value addition, and entrepreneurship development initiatives.
6. *Community Participation and Empowerment*: Promoting the active involvement of tribal communities in decision-making processes, project planning, and implementation. This ensures that interventions

are culturally appropriate, socially inclusive, and responsive to local needs and priorities.

7. *Research and Documentation*: Encouraging research, documentation, and dissemination of best practices and success stories in tribal agriculture and traditional knowledge systems. This helps inform policy formulation, program design, and replication efforts in other tribal regions.

Overall, the "Knowledge Systems and Homestead Agriculture Management in Tribal Areas" initiative aims to foster sustainable development, cultural resilience, and food sovereignty among tribal communities by harnessing their traditional knowledge systems and promoting ecologically sound agricultural practices tailored to local contexts.

Krishi Sakhi Yojana (2023)

The Krishi Sakhi Yojana is a government initiative aimed at empowering women in rural areas by providing them with training and resources to become "Krishi Sakhis" or agricultural friends. Launched on August 30, 2023 with the objective of enhancing women's participation in agriculture and rural development, this scheme equips them with knowledge and skills in various aspects of farming, including crop management, animal husbandry, horticulture, and organic farming.

Key features of the Krishi Sakhi Yojana include

1. *Training and Capacity Building*: Women selected as Krishi Sakhis undergo comprehensive training programs conducted by agricultural experts and extension workers. These training sessions cover various topics related to modern agricultural practices, sustainable farming methods, and use of agricultural machinery and technology.

2. *Extension Services*: Krishi Sakhis serve as a bridge between agricultural experts and rural women farmers. They provide extension services, disseminate information on best agricultural practices, and offer advisory support to women farmers in their communities.

3. *Technology Dissemination*: The scheme aims to promote the adoption of modern agricultural technologies and practices among women farmers. Krishi Sakhis play a crucial role in demonstrating and promoting the use of improved seeds, fertilizers, pesticides, and water-saving irrigation techniques.

4. *Entrepreneurship Development*: Krishi Sakhi Yojana encourages women to explore entrepreneurial opportunities in agriculture and allied sectors. Krishi Sakhis are trained to set up and manage small-scale agribusinesses such as nursery management, vermicomposting, poultry farming, and food processing units.

5. *Financial Support*: The scheme may provide financial assistance, subsidies, or credit facilities to Krishi Sakhis to help them establish their own agricultural enterprises or access inputs and equipment needed for farming activities.

6. *Empowerment and Social Inclusion*: By engaging women in agricultural activities and decision-making processes, the Krishi Sakhi Yojana aims to empower them economically and socially. It enhances their role in household decision-making, improves their bargaining power, and contributes to gender equality and women's empowerment in rural areas.

Overall, the Krishi Sakhi Yojana is a significant initiative that leverages the potential of women as change agents in agriculture and rural development. By equipping them with knowledge, skills, and resources, the scheme not only enhances agricultural productivity but also promotes inclusive and sustainable rural development.

8

New Trends in Agriculture Extension Private Extension

Privatization of Extension or Agriculture Extension Services

The privatization of agricultural extension services refers to the involvement of private entities, such as agribusiness firms, non-governmental organizations (NGOs), and consulting firms, in the delivery of agricultural advisory services traditionally provided by public extension agencies. This shift towards privatization has been observed globally, driven by various factors including fiscal constraints, the perceived inefficiencies of public extension systems, and the increasing role of the private sector in agricultural development.

Concept and Definition

- ***Vanden Ban and Hawkins (1996)*** state that in private sector extension, farmers are expected to share the responsibility for the service and pay all or part of the cost.
- ***Saravanan and Shivalinge Gowda (1999)*** operationalized Privatization as follows: "Privatization of extension service refers to the services rendered in the area of agriculture and allied aspects by extension personnel working in private agencies or organizations for which farmers are expected to pay a fee (or free) and it can be viewed as supplementary or alternative to public extension services".

Global Perspective

Market-Driven Approach: Privatization often reflects a market-driven approach to agricultural extension, where private entities deliver services based on demand and are incentivized by potential profits. This can lead to greater responsiveness to farmers' needs and preferences.

Innovative Technologies: Private extension providers may introduce innovative technologies, practices, and market linkages that can improve agricultural productivity, profitability, and competitiveness. They often leverage their expertise, networks, and resources to deliver tailored solutions to farmers.

Risk of Exclusion: However, there are concerns that privatization may lead to the exclusion of marginalized and small-scale farmers who may not afford or access private extension services. This can exacerbate inequalities in access to information, resources, and markets.

Quality Assurance: Ensuring the quality and credibility of private extension services is essential to maintain farmers' trust and ensure positive outcomes. Regulatory frameworks, certification mechanisms, and standards of practice are needed to uphold the integrity and effectiveness of privatized extension services.

Why Private Extension?

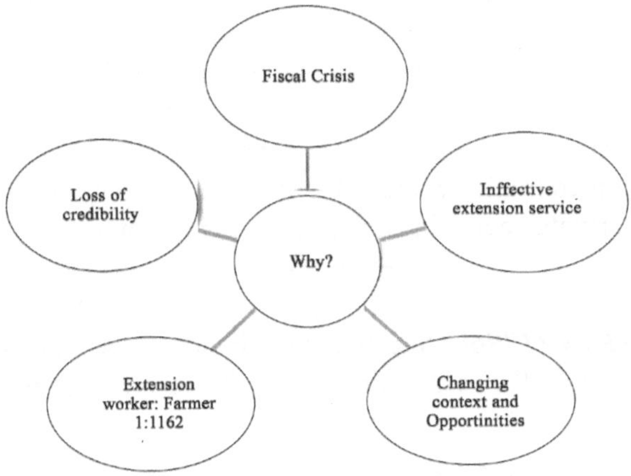

Advantages of Private Extension System

- Reduces the economic burden of government
- Increases the efficiency of extension services
- Increases the accountability of extension agents
- Government can have maximum control over private extension
- Support and control by clients leading to client orientation

Disadvantages of Private Extension System

- It may hamper free flow of information
- Farmers may be less interested to disseminate technologies to other farmers
- Only commercial farmers will be benefited
- Commercial interests of these private agencies may jeopardize the eco-friendly and sustainable approaches
- Contact between farmers and extension agents get declined

Elements of Private Extension

Objectives	• Profit Maximization • Increasing the efficiency
Target Group	• Only those who can pay • More committed and Highly careful • Actively participating
Mandates	• **Technologies:** Location specific, demand driven,timely, profitable • **Input supply:** quality, timely and adequate
Funding	• Clients contribution • Development agencies
Extension Personnel	• Accountable to farmers • Highly motivated • Highly professional
Extension Method	• Personal communication • Participatory approach • Less use of group and mass contact

- Profit Maximization
- Increasing the efficiency

- Only those who can pay
- More committed and Highly careful
- Actively participating

• **Technologies:** Location specific, demand driven,timely, profitable • **Input supply:** quality, timely and adequate
• Clients contribution • Development agencies
• Accountable to farmers • Highly motivated • Highly professional
• Personal communication • Participatory approach • Less use of group and mass contact

Indian Perspective

Government Initiatives: In India, the public sector has historically played a dominant role in agricultural extension through institutions such as the Agricultural Extension Services (ATMA), Krishi Vigyan Kendras (KVKs), and state agricultural universities. However, there has been a growing recognition of the need to involve the private sector in extension service delivery.

Public-Private Partnerships (PPPs): The Indian government has increasingly promoted PPPs in agricultural extension to leverage the expertise, resources, and outreach of both public and private actors. Initiatives such as the Agricultural Technology Management Agency (ATMA) scheme have encouraged collaboration between government agencies, NGOs, and private companies.

Challenges and Opportunities: While privatization offers opportunities for innovation, efficiency, and market orientation, it also poses challenges such as ensuring equity, accountability, and sustainability. Balancing the roles of public and private extension providers while addressing the diverse needs of farmers remains a key challenge in the Indian context.

Role ofNGOs andAgribusiness: NGOs and agribusiness firms have increasingly played a role in delivering extension services in India, particularly in areas such as organic farming, value chain development, and agri-entrepreneurship. They often complement government efforts by focusing on niche areas or underserved communities.

In summary, the privatization of agricultural extension services represents a shift towards a more diversified and market-oriented approach to knowledge dissemination and technology transfer in agriculture. While it offers potential benefits in terms of innovation and efficiency, careful attention must be paid to issues of equity, quality assurance, and regulatory oversight to ensure that the needs of all farmers are addressed and positive outcomes are achieved.

9

New Trends in Agriculture Extension Cyber Extension/ e-Extension

In recent years, the digital revolution has transformed various sectors, including agriculture. Cyber Extension refers to the use of Information and Communication Technologies (ICTs) to deliver extension services and agricultural information to farmers. Let's delve deeper into this concept.

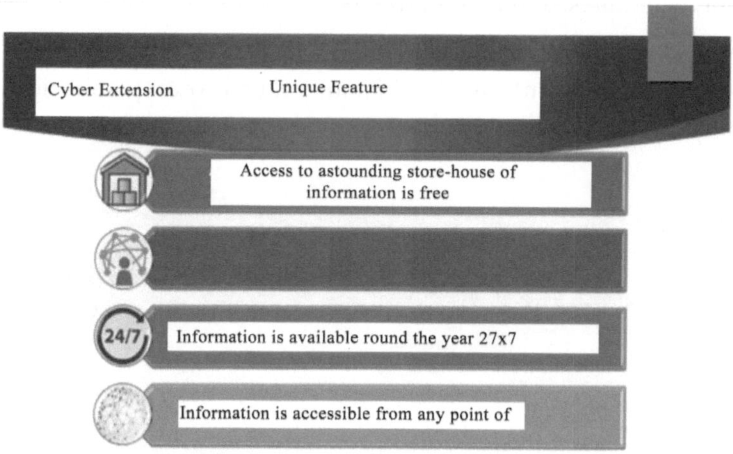

1. Evolution of Extension Services

- Traditional extension services involved face-to-face interactions between extension agents and farmers. While effective, these services often faced challenges such as limited reach, high costs, and scalability issues.
- With the advent of ICTs, extension services have undergone a significant transformation, leading to the emergence of Cyber Extension.

2. Key Components of Cyber Extension

- *Digital Platforms:* Cyber Extension utilizes various digital platforms such as websites, mobile applications, social media, and online forums to disseminate agricultural information and interact with farmers.

- *Remote Sensing and GIS:* Technologies like remote sensing and Geographic Information Systems (GIS) are employed to provide farmers with real-time data on weather, soil moisture, crop health, and pest outbreaks.
- *ICT Tools:* Mobile phones, tablets, and other ICT tools enable farmers to access extension services, receive advisory messages, and participate in virtual training sessions from anywhere, at any time.
- *Mobile Technology:* Mobile phones and text messaging services enable farmers to receive advisory messages, weather alerts, market prices, and other relevant information directly on their devices.
- *E-Learning:* Cyber Extension facilitates e-learning modules, webinars, and online courses to educate farmers on best agricultural practices, crop management techniques, and market trends.
- *Decision Support Systems:* Decision support tools and expert systems assist farmers in making informed decisions related to crop selection, irrigation scheduling, fertilizer application, and pest management.

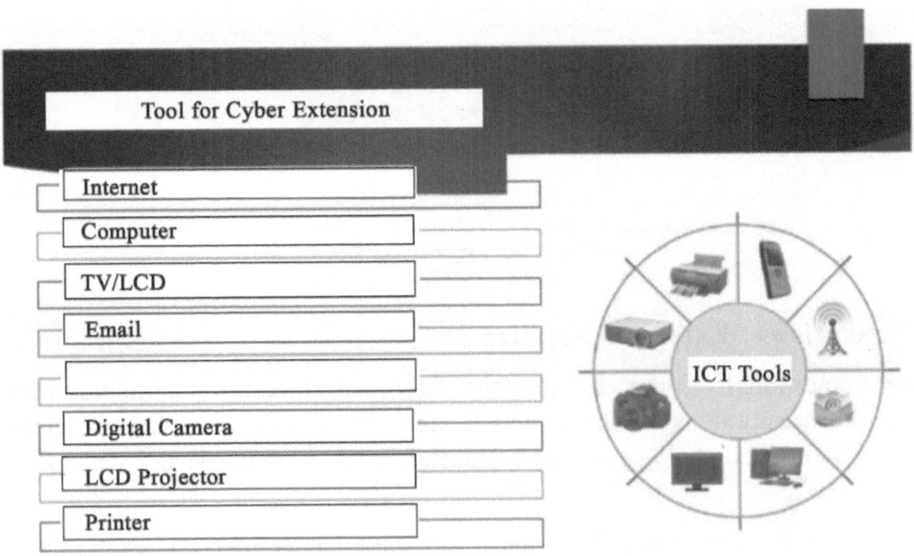

3. Benefits of Cyber Extension

- *Enhanced Outreach:* Cyber Extension enables extension services to reach a wider audience, including remote and marginalized communities, thereby promoting inclusivity.

- *Timely and Relevant Information:* Farmers receive timely and relevant information on weather forecasts, market prices, government schemes, and agronomic practices, enhancing their decision-making capabilities.
- *Cost-Effectiveness:* Compared to traditional extension methods, Cyber Extension offers cost-effective solutions, reducing the need for extensive travel and infrastructure investments.
- *Empowerment of Farmers:* By providing access to knowledge and resources, Cyber Extension empowers farmers to adopt modern technologies, increase productivity, and improve their livelihoods.

4. Challenges and Considerations

- *Digital Divide:* Disparities in access to technology, internet connectivity, and digital literacy may exacerbate inequalities, limiting the effectiveness of Cyber Extension in reaching all farmers.
- *Content Localization:* Ensuring that agricultural information is culturally appropriate, locally relevant, and available in regional languages is crucial for effective communication with farmers.
- *Data Security and Privacy:* Safeguarding farmers' data and privacy rights is essential to build trust and credibility in Cyber Extension initiatives.

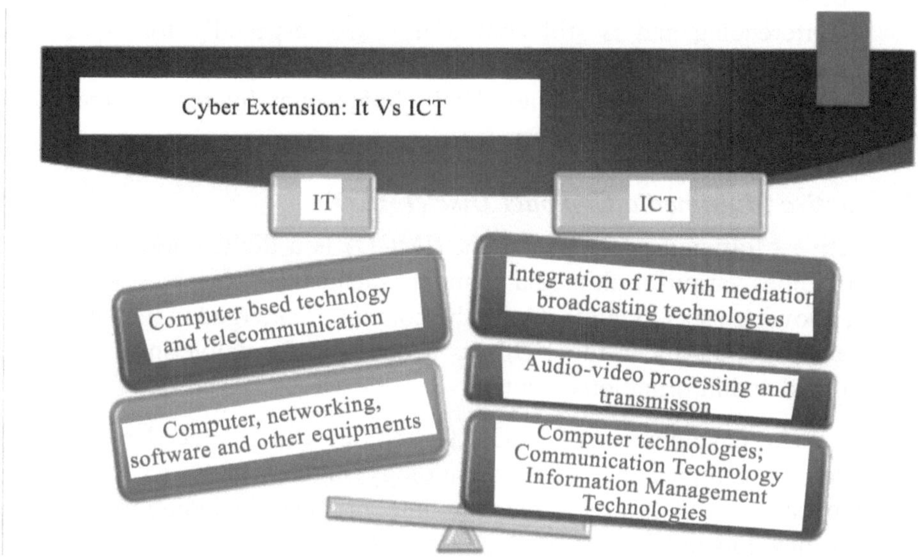

Tools of ICT

(i) *Internet:* The Internet is a global network of computers that enables the exchange of information and communication through various services such as email, websites, social media, and more. It has revolutionized how people connect, access information, and conduct business worldwide.

(ii) *Cyber Cafes*: Cyber cafes, also known as internet cafes, are establishments where individuals can pay to access the internet on computers provided by the cafe. They were popular in the late 1990s and early 2000s when home internet access was not as widespread. Although their popularity has declined with the widespread availability of internet access on personal devices, they still serve as convenient options for travelers or those without personal internet access.

(iii) *Video Conferencing*: Video conferencing allows individuals or groups in different locations to communicate in real-time through audio and video transmission. It has become increasingly popular for remote work, business meetings, online education, and socializing. Platforms like Zoom, Microsoft Teams, and Google Meet have seen a surge in usage, particularly during the COVID-19 pandemic, when remote communication became essential.

(iv) *Teleconferencing*: Teleconferencing refers to audio-based communication between multiple parties in different locations. It predates video conferencing and is still commonly used, especially for audio-only meetings or conference calls. Teleconferencing can be conducted over traditional phone lines, VoIP (Voice over Internet Protocol), or specialized conferencing services.

Interactive Multimedia Compact Disc (IMCD)

Interactive Multimedia Compact Disk (IMCD) is a digital storage medium that combines various multimedia elements such as text, graphics, audio, and video, allowing users to interact with the content. It emerged in the late 1980s and gained popularity in the 1990s as a platform for delivering interactive educational and informational content.

History: The development of IMCD technology can be traced back to the evolution of compact disc (CD) technology in the 1980s. As CD-ROMs became capable of storing large amounts of data, including multimedia content, developers began to explore interactive applications beyond simple text-based storage. This led to the creation of IMCDs, which could deliver engaging multimedia experiences to users.

Revolution in Agriculture: In the field of agriculture, IMCDs have revolutionized education, training, and information dissemination. They have been used to create interactive tutorials, training programs, and informational resources covering various aspects of agriculture, including crop cultivation techniques, pest management, livestock farming, agricultural machinery operation, and more.

1. *Education and Training:* IMCDs have been instrumental in providing interactive educational materials to farmers, agricultural students, and professionals. These multimedia resources offer visual demonstrations, simulations, and interactive exercises to enhance learning outcomes. Farmers can access valuable information on best practices, crop management techniques, and innovative farming methods, improving their productivity and efficiency.
2. *Information Dissemination:* IMCDs serve as a valuable tool for disseminating agricultural information to rural communities and remote areas with limited access to traditional educational resources. They can deliver localized content in multiple languages, catering to the specific needs and challenges of different regions. This helps bridge the knowledge gap and empowers farmers with relevant information to make informed decisions.
3. *Extension Services:* Agricultural extension services utilize IMCDs to provide on-demand access to expert advice, market information, weather forecasts, and other relevant resources. Farmers can benefit from personalized guidance and recommendations tailored to their specific agricultural practices and local conditions, improving their decision-making process and overall farm management.

Overall, IMCDs have played a significant role in modernizing agriculture, promoting sustainable practices, and empowering farmers with knowledge and tools to overcome challenges and enhance productivity in a rapidly evolving agricultural landscape.

Agri-Portals

A portal is a website or online platform that serves as a gateway to a variety of information, services, and resources related to a specific topic or domain. Portals typically offer a centralized hub where users can access a wide range of content, tools, and functionalities tailored to their needs. In the realm of agriculture, there are numerous portals that provide farmers, researchers, policymakers, and other stakeholders with valuable information, services, and support. These portals cover various aspects of agriculture, including crop cultivation, livestock management, market information, weather forecasts, government schemes, research findings, and more. One such example is the Agritech Portal of Tamil Nadu Agricultural University (TNAU). TNAU is a leading agricultural university in India, known for its research and extension activities aimed at promoting agricultural development and innovation.

Agritech Portal of TNAU: This portal is an online platform developed by Tamil Nadu Agricultural University to provide farmers with access to agricultural information, technologies, and services. Here's a brief overview of its features:

1. *Crop Information:* The portal offers comprehensive information on crop cultivation practices, including crop varieties, planting techniques, nutrient management, pest and disease control, and harvesting methods. Farmers can access detailed guidelines and recommendations for various crops grown in Tamil Nadu.
2. *Weather Information:* Weather plays a crucial role in agricultural planning and decision-making. The portal provides real-time weather updates, forecasts, and agro-advisories to help farmers anticipate weather patterns and take appropriate actions to mitigate risks and optimize farm operations.
3. *Market Intelligence:* Access to market information is essential for farmers to make informed decisions about crop marketing and pricing. The portal provides market intelligence, including commodity prices, market trends, demand-supply dynamics, and marketing strategies to help farmers maximize their profits.
4. *Extension Services:* TNAU's Agritech Portal serves as a platform for delivering extension services to farmers. It offers interactive features such as forums, helplines, and expert consultations, enabling farmers to seek guidance, ask questions, and exchange knowledge with agricultural experts and fellow farmers.
5. *Technology Adoption:* The portal promotes the adoption of modern agricultural technologies and practices by providing information about innovative farming techniques, farm machinery, irrigation methods, and sustainable agriculture initiatives.

(Source: https://agritech.tnau.ac.in/)

Overall, the Agritech Portal of Tamil Nadu Agricultural University serves as a valuable resource for farmers in Tamil Nadu, empowering them with knowledge, tools, and support to enhance agricultural productivity, profitability, and sustainability.

Information Kiosks

Information kiosks are interactive computer terminals or booths that provide access to various types of information, services, and applications to users in public places such as airports, train stations, shopping malls, libraries, and government offices. These kiosks typically feature a touchscreen interface and may also include input devices such as keyboards, mice, or trackballs for user interaction.

Features of Information Kiosks

1. *Accessibility*: Information kiosks are designed to make information readily accessible to the public, allowing users to retrieve information on-demand without the need for assistance from staff members.

2. *Multimedia Content*: Kiosks often offer multimedia content such as text, images, videos, and audio to provide users with a rich and engaging experience. This can include maps, directories, instructional videos, promotional materials, and more.
3. *Self-Service Functions*: Information kiosks enable users to perform self-service functions such as checking flight schedules, locating stores or services within a facility, accessing public transportation information, paying bills, printing documents, and completing transactions.
4. *Convenience*: By placing information at users' fingertips, information kiosks enhance convenience and efficiency, reducing the need for users to wait in lines or seek assistance from staff members.
5. *24/7 Availability*: Many information kiosks are available 24/7, providing users with access to information and services outside of regular business hours.
6. *Customization*: Information kiosks can be customized to suit the specific needs of different locations and user groups. They can be programmed to display information in multiple languages, offer accessibility features for users with disabilities, and provide tailored content based on user preferences or demographics.

Information kiosks serve as valuable tools for disseminating information, providing services, and enhancing user experiences in various public settings. They offer a convenient and efficient way for users to access information and complete tasks independently.

Kisan Call Centre

The Department of Agriculture & Cooperation (DAC), Ministry of Agriculture, Govt. of India launched Kisan Call Centers on January 21, 2004 across the country to deliver extension services to the farming community. The purpose of these call centers is to respond to issues raised by farmers, instantly, in the local language. There are call centers for every state which are expected to handle traffic from any part of the country. Queries related to agriculture and allied sectors are being addressed through these call centres.

The Department of Agriculture & Cooperation, Ministry of Agriculture, Govt. of India has launched Kisan Call Centers with a view to leverage the extensive telecom infrastructure in the country to deliver extension services to the farming community. The purpose of these Call Centers is mainly to respond to issues raised by farmers instantly in the local language, on continuous basis.

The country today has an impressive telecom network both in the private and Government sector. Over 5 lakh villages have a public telephone in the country. It has been felt for long that this impressive telecom network could be put to effective use for delivering knowledge and information to the farming community.

A call centre based extension service will be delivering knowledge and information exactly as per the requirements of the farming community. This system would also help keep a record of what is being delivered to the farmers in terms of knowledge and information. The Kisan Call Centre scheme is available over the complete country. The Kisan Call Centre scheme has been functioning from 21.1.04. The Call Centres can be accessed by farmers all over the country on common Toll Free Number 1551.

Since 10th June, 2004 the Call Centres service has been made available right from 6 A.M. to 10 P.M. except on Sundays and gazetted holidays, beyond these hours the calls are attended in the IVRS mode. The call centre for the State of Gujarat and U.T. of Dadra and Nagar Haveli has started functioning effectively w.e.f. 1.11.2004 at Ahmedabad. The calls are received at 13 Call Centres wherein 116 Agriculture Graduates attend to answer the queries of the farmer in the local language. 123 experts located in different parts of the country at State Agriculture Universities, ICAR institutes, State Department of Agriculture, Horticulture and other developments are answering the calls at Level –II.

The Kisan Call Center infrastructure is placed at three locations namely

(1) A professionally managed Call Center (Level-I)

(2) A Response Center in each organization, where services of Subject Matter Specialists are made available (Level-II)

(3) The Nodal Cell (Level-III)

The Kisan Call Centre scheme has been functioning from 21.1.04. The Call Centres can be accessed by farmers all over the country on common Toll Free Number 1800-180-1551.

Mobile phone and Telephone Technology

Telephone and mobile phones have established itself in the rural areas and are becoming quite popular both with farmer and farm women. These powerful electronic machines that were a farmer's dream earlier have become a reality as the farmers can immediately make use of them to address their field problems and other farm difficulties. The government gifted Kisan Call Centres (KCC) are functioning all over India which give answers to the farmer queries in local languages.

For example: Grameen Phone is a commercial operation providing cellular services in both urban and rural areas of Bangladesh, with approximately 40,000 customers. A pilot programme of Grameen Phone, through the Grameen Bank and a wholly owned subsidiary called Grameen Telecom, is enabling women members of the Grameen Bank's revolving credit system to retail cellular phone services in rural areas. This instrument is more than a means of communication: it is being used as a weapon for empowerment and fight poverty.

SMS (Short Message Service)

Short Message Service (SMS) is the text communication service component of phone, web or mobile communication systems, using standardized communications protocols that allow the exchange of short text messages between fixed line or mobile phone devices. SMS text messaging is the most widely used data application in the world, with 2.4 billion active users, or 74% of all mobile phone subscribers The term SMS is used as a synonym for all types of short text messaging as well as the user activity itself in many parts of the world.

SMS has used on modern handsets was originated from radio telegraphy in radio memo pagers using standardized phone protocols and later defined as part of the Global System for Mobile Communications (GSM) series of standards in 1985 as a means of sending messages of up to 160 characters], to and from GSM mobile handsets. Since then, support for the service has expanded to include other mobile technologies such as ANSI CDMA networks and Digital AMPS, as well as satellite and landline networks. Most SMS messages are mobile-to-mobile text messages though the standard supports other types of broadcast messaging as well.

Some of the Indian Experiences on Information and Communication Technology

- LOKMITRA Project at Himachal Pradesh
- Warana Wired Village Project, Maharashtra
- GYANDOOT Project at Dhar, Madhya Pradesh
- TARAHAAT
- JAN MITRA Project at Rajasthan

Village Knowledge CENTRE: (VKC)

The Village Knowledge Centre (VKC) is a place to render distant services from a single window point to rural masses especially in remote areas of the country through modern Information and Communication Technology. villages Information Center that enable rural families access a basket of modern information, internet connectivity through dial up telephones and communication technologies. Training educated youth, especially women, in rural areas in operating information center.

The mission of the VKCs is to harness the power of information and communication technologies in the knowledge, skill, economic and social empowerment of rural families. With the kind of awareness and opportunities available in ICT, it will soon become a reality wherein every one of our villages will have computers and connectivity available. These would be the window to the world of knowledge for our villages and also to reap the benefits of our e-governance, tele-education, tele-medicine, e-commerce and e-judiciary initiatives.

To empower the villagers with the knowledge and to act as a nodal center for knowledge connectivity for the villagers. The knowledge center from which the villagers would access the information through the village information officer can also be used for collection, digital storage and dissemination of village specific information pertaining to any relevant information to the villagers.

Objectives

1. To inculcate reading habits among the children through our library in VKC.
2. To make the children familiar with computers.
3. To connect the villagers to the external world through newspapers, magazines, computers, transistor and television that are set up in our VKC

The M.S. Swaminathan Research Foundation (MSSRF) is a non-profit organization founded by the noted Indian Food and Agricultural scientist, Dr. M.S. Swaminathan. The foundation was registered in 1988 to research, advance and promote coastal systems, biotechnology, biodiversity, eco technology, food security and information, education and communication in developing countries, with a specific focus on India. In 1998, the foundation started the "Village Knowledge Centers" project. The idea was to select villages in rural Tamil Nadu and the Union Territory of Pondicherry (both in southern India), and provide adequate telecommunications infrastructure that would enable the dissemination of appropriate information regarding farming, education, health, weather, governmental news, job, loans and aid opportunities.

The objectives were to reduce the digital gap and gender divide in rural India using technology – especially telecommunications technology. Each Village Knowledge Center (VKC) would serve one or more villages and would act as the knowledge repository for the villages it served. Pertinent information was stored in databases in these sites, with appropriate technology for easy access and dissemination of this knowledge to persons seeking them. The long term objective was to eventually set up VKCs in each of the approximately 638, 000 villages in India, thus creating a nation-wide rural knowledge repository.

Geographical Information System (GIS)

Geographical Information System (GIS) is a technology that provides the means to collect and use geographic data to assist in the development of Agriculture. A digital map is generally of much greater value than the same map printed on a paper as the digital version can be combined with other sources of data for analyzing information with a graphical presentation. The GIS software makes it possible to synthesize large amounts of different data, combining different layers of information to manage and retrieve the data in a more useful manner. GIS provides a powerful means for agricultural scientists to better service to the farmers and farming community in answering their query and helping in a better decision making to implement planning activities for the development of agriculture.

E-choupals

Traditionally, choupals are community gathering places in the village where locals meet to discuss issues and iron out their problems. In the digital age, e-choupals are gradually revolutionizing the way Indian farmers do business. The concept of e-choupal was pioneered by one of India's largest exporters of agricultural commodities, Indian Tobacco Company's International Business Division (ITC_IBD). E-choupals constituted on internet enabled kiosk in a village which is manned by a prominent local farmer who is familiar with computers, known as the 'choupal sanchalak'. The setting up of each e-choupal entails as investment of between Rs. 1 to 3 lacs.

The sanchalak man of the kiosk is in touch with ITC representatives and guides farmers on the use of the technology. Given the levels of literacy and infrastructure limitations the sanchalak acts as the interface between the computer and the farmer. Farmers can use the kiosks to check the current market prices of their commodities, access market data, get information on local and global weather and best farming practices. Presently, near about 2,500 internet kiosks were established in villages by ITC across 18 states of the country. The total investment was near about Rs. 1500 crores. These technologies have transformed communication, enabling people to connect and collaborate across geographical boundaries with ease.

In conclusion, Cyber Extension holds immense potential to revolutionize agricultural extension services and empower farmers with knowledge, resources, and decision support tools. However, addressing challenges such as the digital divide and ensuring content localization and data security are imperative for maximizing its impact and ensuring inclusivity in agricultural development.

10

New Trends in Agriculture Extension Farmer-Led Extension

Concept

The concept of Farmer-Led Extension (FLE) embodies a participatory approach wherein farmers actively engage in sharing their knowledge, experiences, and best practices with their peers within the community. This approach harnesses the credibility and trust that successful farmers have established among their fellow farmers. The selection criteria for farmer extensionists prioritize attributes such as innovativeness, diligence, honesty, credibility, community acceptance, and a willingness to both learn and impart knowledge. A farmer extensionist, ideally, should have a track record of implementing innovative farming techniques, actively participating in agricultural trials and field demonstrations, and preferably attending Farmer Field Schools (FFS). Moreover, the farmer extensionist should demonstrate the financial capacity and willingness to invest in adopting new agricultural technologies. The Farmer-Led Extension model encompasses various modalities, including:

1. *Farmers-to-Farmer:* This involves farmers providing training and knowledge transfer directly to their peers, fostering a peer-to-peer learning environment.
2. *Farmer Field School:* A structured approach where groups of farmers collectively engage in hands-on learning activities in the field, facilitated by experienced farmer trainers.
3. *The Problem Census:* Farmers collaboratively identify and prioritize agricultural challenges within their community, laying the groundwork for targeted interventions and solutions.
4. *Problem-Solving Approach:* Encourages farmers to collectively tackle agronomic challenges through a systematic problem-solving methodology, fostering collaboration and shared learning.
5. *NGO-Government Collaboration:* Partnerships between non-governmental organizations (NGOs) and government agencies leverage

resources and expertise to support farmer-led extension initiatives, enhancing outreach and impact.

In essence, Farmer-Led Extension empowers farmers to take ownership of their agricultural development by leveraging their local knowledge, expertise, and social networks to disseminate best practices, address challenges, and foster sustainable agricultural growth within their communities.

Paradigm shift from production-led to farmers-led extension system (Kokate *et al.*, 2009)

Components	Production-led	Farmers-led
Purpose/ Objective	Transfer of production technologies	Capacity building (especially farmers extensionist), create para-professional technologies extension workers, creating or Strengthening local institution
Goal	Food self-sufficiency	Livelihood security including food, nutrition, employment to alleviate poverty Sustainability and conserving bio-diversity
Approach	Top-down, commodity and supply driven	Participatory, bottom-up and demand driven
Actors	Mostly public institutions	Pluralistic with public, private, non-government and farmers organizations as a partner rather than competitors
Mode	Mostly interpersonal/ individual approach	Integration of clients oriented on-farm participatory/ experiential learning methods supported by ICTs and media
Role of extensionagents	Limited to delivery mode and feedback to research system	Facilitation of learning, building overall capacity of farmers and encouragingfarmers experimentation
Linkages/ liaison	Research-Extension-Farmers	Research-Extension-Farmers Organizations (FIGs, CIGs, SHGs)
Emphasis	Information management, Production "Seed to Seed"	Knowledge management and sharing
Nature of technology	Input intensive, crop based andgeneral recommendations as per agro-climatic zone, fixed package of information	Knowledge intensive, broad based, farming system perspective and blending with ITKs
Critical areas	Improvement, production and protection	Decision support system, integrated farming system approach, natural resource management, client group formation and Community empowerment
Critical inputs	Money and material	Access to information, building human and social capital
Accountability	Mostly government	To farmers rather than donors

11

New Trends in Agriculture Extension Market-Led Extension

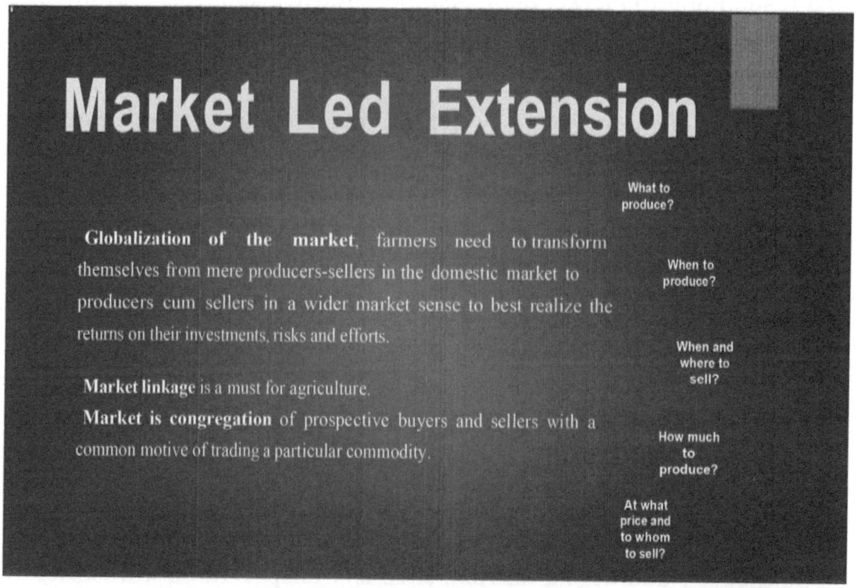

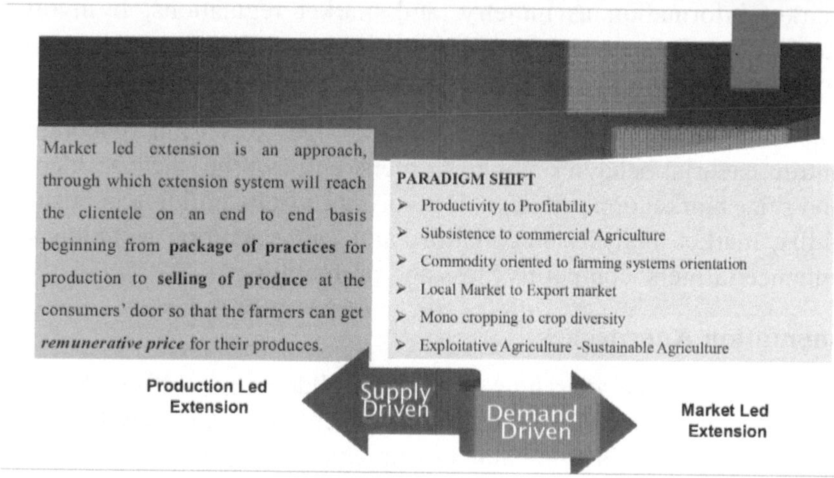

Market-led extension represents a paradigm shift in agricultural extension, focusing on market dynamics and opportunities to drive agricultural development. Unlike traditional extension approaches, which primarily emphasize technology transfer, market-led extension prioritizes market-oriented interventions to enhance farmers' profitability, competitiveness, and resilience.

Key Concepts:

1. *Market Orientation:* Market-led extension emphasizes understanding and responding to market demands, preferences, and trends. It involves analyzing market dynamics, identifying profitable market opportunities, and aligning agricultural production with market requirements.
2. *Value Chain Development:* A central tenet of market-led extension is value chain development, which involves strengthening linkages between farmers, input suppliers, processors, traders, and consumers. Value chain interventions aim to enhance efficiency, quality, and value addition at each stage of the agricultural value chain.
3. *Market Information Services:* Access to timely and accurate market information is crucial for farmers to make informed decisions regarding crop selection, production practices, pricing, and marketing strategies. Market-led extension facilitates the dissemination of market information through various channels, including mobile phones, radio, internet platforms, and market intelligence systems.
4. *Market Access:* Market-led extension seeks to improve farmers' access to markets by addressing barriers such as transportation, infrastructure, market information asymmetry, and market regulations. It involves supporting farmers in establishing market linkages, negotiating contracts, and complying with quality and certification standards.
5. *Entrepreneurship Development:* Market-led extension encourages entrepreneurial behavior among farmers, empowering them to identify and seize market opportunities. It involves capacity building in business skills, market analysis, negotiation, and value addition techniques to enhance farmers' competitiveness and profitability.

Implementation Approaches

1. *Public-Private Partnerships (PPPs):* Collaborative efforts between government agencies, private sector entities, and civil society organizations are instrumental in implementing market-led extension

initiatives. PPPs leverage the strengths and resources of each partner to deliver targeted extension services, market interventions, and infrastructure development.

2. *Value Chain Development Programs:* Integrated value chain development programs focus on enhancing the efficiency, competitiveness, and inclusivity of agricultural value chains. These programs involve stakeholders across the value chain in participatory planning, capacity building, and investment in infrastructure and technology.
3. *ICT-Based Solutions:* Information and Communication Technology (ICT) tools such as mobile applications, e-commerce platforms, and digital extension services play a critical role in facilitating market-led extension. ICT-based solutions provide farmers with access to market information, extension advice, financial services, and market linkages.
4. *Capacity Building:* Capacity building initiatives are essential for equipping extension agents, farmers, and other value chain actors with the knowledge, skills, and tools required for market-led extension. Training programs cover topics such as market analysis, value chain management, post-harvest handling, and business planning.

Benefits and Challenges

1. **Benefits**
   - Increased farmer income and profitability
   - Enhanced market access and competitiveness
   - Improved agricultural productivity and quality
   - Diversification of agricultural products and markets
   - Strengthened resilience to market fluctuations and shocks
2. **Challenges**
   - Limited access to market information and infrastructure in remote areas
   - Capacity gaps among farmers and extension agents in market analysis and value chain management
   - Market volatility and price fluctuations affecting farmer income and investment decisions
   - Institutional barriers and policy constraints hindering market access and value chain development

- Sustainability of market-led extension initiatives beyond project timelines and funding cycles

Market-led extension represents a transformative approach to agricultural development, placing markets at the center of extension interventions. By empowering farmers to respond to market opportunities and challenges, market-led extension contributes to inclusive economic growth, poverty reduction, and sustainable agricultural development.

12

New Trends in Agriculture Extension Expert Systems

Expert System

An Expert System is a software application designed to replicate the problem-solving abilities of one or more human experts within a specific domain. It utilizes knowledge and inference procedures to analyze complex problems, provide expert advice, and make decisions.

> ***Definition***
>
> Expert system is an intelligent computer program that uses knowledge and inferences procedures to solve problems.
>
> ***-Daniel Hunt (1986)***
>
> Expert system can be defined as a tool for information generation from knowledge. It is a computer programme designed to stimulate the problem-solving behaviour of an expert in a narrow domain or discipline.
>
> ***-Prasad and Babu (2008)***

Origins in Artificial Intelligence (AI)

The concept of Expert Systems emerged from the field of Artificial Intelligence (AI), which seeks to develop intelligent machines capable of performing tasks that typically require human intelligence. Expert Systems represent a specialized branch of AI focused on capturing and utilizing human expertise in specific domains.

Key Components

1. *Knowledge Base:* The Knowledge Base serves as the repository of domain-specific information, rules, heuristics, and procedures encoded in the system. It encapsulates the expertise of human experts and provides the foundation for problem-solving.

2. *Inference Engine:* The Inference Engine is the reasoning component of the Expert System responsible for applying logical rules and algorithms to deduce conclusions from the information stored in the Knowledge Base. It simulates human reasoning processes to analyze problems and generate solutions.
3. *User Interface:* The User Interface facilitates interaction between the Expert System and users, allowing them to input queries, receive advice, and interpret results. It can vary from simple text-based interfaces to more sophisticated graphical interfaces depending on the application requirements.

Functionality

Expert Systems operate by employing a set of predefined rules and algorithms to process input data, assess the current problem scenario, and generate recommendations or solutions. They use various reasoning methods such as deduction, induction, and abduction to draw logical inferences and reach conclusions.

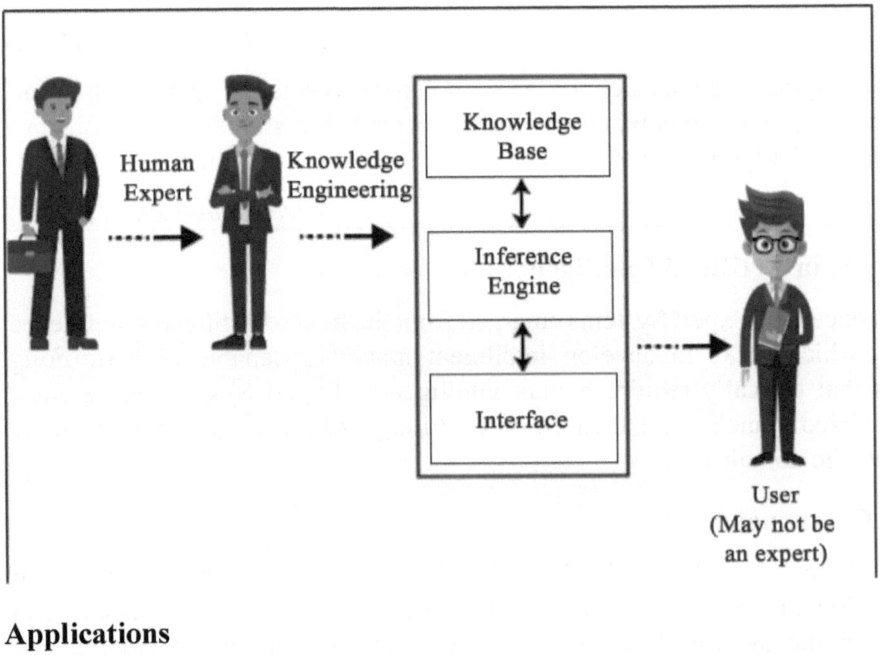

Applications

Expert Systems find applications across diverse domains, including but not limited to:

- *Medical Diagnosis:* Assisting healthcare professionals in diagnosing diseases and recommending treatment options based on symptoms and patient data.
- *Financial Analysis:* Providing investment advice, risk assessment, and portfolio management strategies to investors and financial analysts.
- *Engineering Design:* Supporting engineers in designing complex systems, troubleshooting equipment failures, and optimizing performance parameters.
- *Customer Support:* Offering personalized recommendations, troubleshooting assistance, and problem-solving guidance to customers in various industries.

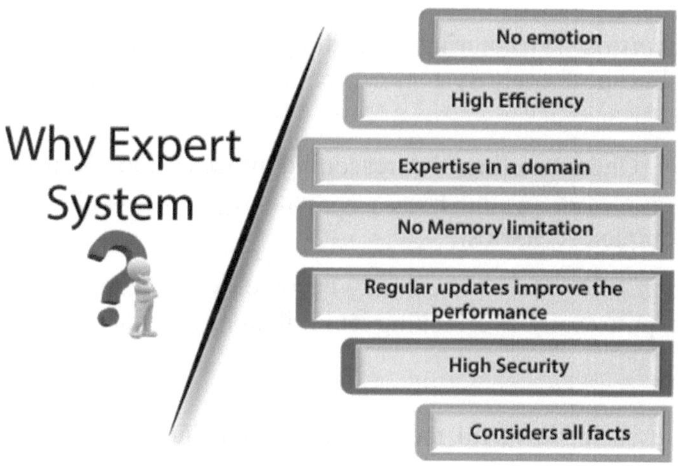

Difference between Conventional Extension and Expert System Extension

(Bahal *et al.*, 2004)

S. No.	Conventional Extension	Expert System of Extension
1.	Universal approachability of same information is a problem	Universal approachability of same information is possible
2.	Information is given whatever is available without consideringneeds and resources	Information is chosen based on their needs and resources
3.	No Cost benefit analysis	Cost benefit analysis
4.	Information flow depends on availability of agent	Information through Cyber Cafe at any place at any time
5.	Require users to draw their own conclusion from facts	Conclusion is drawn based on thedecision given by the expert

Advantages

- *Access to specialized expertise:* Expert Systems provide access to domain-specific knowledge and expertise that may not be readily available or easily accessible.
- *Consistency and reliability:* They ensure consistent decision-making and problem-solving outcomes based on predefined rules and logical algorithms.
- *Scalability and flexibility:* Expert Systems can be easily updated, expanded, and customized to accommodate changes in domain knowledge and problem requirements.

Challenges

- *Knowledge acquisition:* Capturing, encoding, and updating domain knowledge within the Expert System can be labor-intensive and time-consuming.
- *Interpretability:* Understanding the reasoning process and decision-making logic employed by the Expert System may pose challenges, especially in complex scenarios.
- *Overreliance and trust issues:* Users may become overly reliant on the Expert System, potentially overlooking its limitations and biases.

Conclusion

Expert Systems represent a powerful tool for harnessing human expertise, augmenting decision-making processes, and addressing complex problems in diverse domains. By combining AI techniques with domain-specific knowledge, Expert Systems offer valuable insights, recommendations, and solutions to users, contributing to enhanced efficiency, productivity, and innovation.

13

Rural Development

Meaning, Definitions and Concept

Rural development is a multifaceted process aimed at improving the economic and social well-being of populations residing in rural areas. It encompasses various strategies and initiatives geared towards enhancing the quality of life for rural communities, particularly those belonging to low-income groups. To grasp the concept of rural development fully, it is essential to understand the intertwined concepts of "rural" and "development."

Defining "Rural"

The term "rural" typically refers to areas characterized by low population density and small settlements. It denotes regions where agricultural activities often dominate the economy and where communities may face unique challenges due to geographical isolation or limited access to resources and services.

Understanding "Development"

In the context of rural development, "development" encompasses a process aimed at improving the standards of living for the rural population, particularly those with low incomes. It involves initiatives geared towards economic growth, social progress, and overall improvement in the well-being of rural communities.

> ***In the words of Robert Chambers (1983)***, Rural development is a strategy to enable a specific group of people, poor rural women and men, to gain for themselves and their children more of what they want and need. It involves helping the poorest among those who seek a livelihood in the rural areas to demand and control more of the benefits of rural development. The group includes small-scale farmers, tenants and landless.

Concept of Rural Development

Rural development is not merely a set of isolated interventions but rather a comprehensive strategy aimed at fostering sustainable progress in rural areas.

It focuses on addressing the needs of marginalized groups such as smallholder farmers, tenants, and landless individuals. The ultimate goal of rural development is to uplift the living standards of the rural poor while ensuring the sustainability of the development process.

Dimensions of Rural Development

Conceptual Dimension: Rural development involves the conceptualization of strategies and approaches to foster positive change in rural areas, considering factors such as economic opportunities, social equity, and environmental sustainability.

Phenomenal Dimension: It encompasses the dynamic interactions between various institutional factors, including government agencies, non-governmental organizations, and community-based organizations, in driving rural development initiatives forward.

Strategic Dimension: Rural development serves as a strategic approach to addressing the multifaceted challenges faced by rural communities, including poverty, food insecurity, inadequate infrastructure, and limited access to essential services.

Objectives of Rural Development

The primary objective of rural development is to improve the quality of life for rural populations, particularly those who are economically disadvantaged and socially marginalized. Key objectives include:

1. Enhancing livelihood opportunities for rural communities
2. Strengthening agricultural productivity and food security
3. Promoting equitable access to essential services such as healthcare, education, and infrastructure
4. Fostering social inclusion and empowerment of marginalized groups
5. Facilitating sustainable natural resource management and environmental conservation

Democratic Decentralization

Democratic decentralization refers to the transfer of political power, administrative authority, and decision-making processes from central governments to local governments or grassroots-level institutions. It aims to empower local communities, enhance participation in governance, and promote accountability and responsiveness to local needs.

In India, democratic decentralization came into prominence with the adoption of the 73rd and 74th Constitutional Amendments in 1992. These amendments mandated the establishment of Panchayati Raj institutions (PRIs) at the village, intermediate (block), and district levels for rural areas, and urban local bodies (ULBs) for urban areas. The primary objective was to decentralize power and resources, promote local self-governance, and strengthen democracy at the grassroots level.

The 73rd Amendment Act, 1992, introduced democratic decentralization in India, followed by the 74th Amendment Act, 1992, which focused on urban local bodies. These amendments were inspired by the principles of decentralization outlined in the Constitution of India and recommendations of various committees and commissions, including the Balwant Rai Mehta Committee (1957) and the Ashok Mehta Committee (1977).

Recommendations of Balwant Rai Mehta Committee (1957)

The Balwant Rai Mehta Committee, officially known as the "Committee on Panchayati Raj Institutions," was appointed by the Government of India in 1957 to study the functioning of the community development program and recommend measures for strengthening democratic decentralization in India. The committee's recommendations laid the foundation for the establishment of Panchayati Raj institutions (PRIs) in the country. Some of the key recommendations of the Balwant Rai Mehta Committee are as follows:

1. *Three-Tier System:* The committee recommended the establishment of a three-tier system of Panchayati Raj institutions at the village, intermediate (block), and district levels. This tiered structure aimed to ensure effective representation and administration at different levels of local governance.

2. *Democratic Elections:* The committee advocated for the conduct of regular and democratic elections to Panchayati Raj institutions to ensure the participation of local communities in decision-making processes. It emphasized the principles of universal adult suffrage and direct elections for electing representatives at all levels.

3. *Devolution of Powers:* The committee proposed the devolution of administrative, financial, and planning powers to Panchayati Raj institutions to enable them to function as autonomous and self-governing bodies. It recommended the transfer of responsibilities related to local planning, implementation of development programs, and management of local resources to PRIs.

4. *Financial Resources:* The committee stressed the importance of adequate financial resources for Panchayati Raj institutions to effectively discharge their functions. It recommended the allocation of funds from both central and state governments through statutory provisions to ensure financial autonomy and stability for PRIs.
5. *Role of State Governments:* The committee highlighted the need for supportive legislation by state governments to institutionalize Panchayati Raj institutions and define their powers, functions, and responsibilities. It recommended the enactment of state-level Panchayati Raj Acts to provide a legal framework for local governance.
6. *Capacity Building:* The committee emphasized the importance of capacity building and training programs for elected representatives and functionaries of Panchayati Raj institutions to enhance their skills, knowledge, and understanding of local governance processes.
7. *Role of District Administration:* The committee recommended the involvement of district administration in facilitating and coordinating the functioning of Panchayati Raj institutions, especially in areas related to planning, supervision, and monitoring of development activities.

Overall, the recommendations of the Balwant Rai Mehta Committee laid the groundwork for the decentralization of power and resources to grassroots-level institutions in India, ultimately leading to the enactment of the 73rd Constitutional Amendment Act in 1992, which mandated the establishment of Panchayati Raj institutions across the country.

Three tier system of Panchayat Raj

The three-tier system of Panchayati Raj refers to the organizational structure of local self-government institutions at three levels: the village level (Gram Panchayat), the intermediate level (Panchayat Samiti or Mandal Parishad), and the district level (Zila Parishad). This system was envisaged to promote democratic decentralization and grassroots governance in India.

1. Gram Panchayat (Village Level)

- The Gram Panchayat is the lowest level of Panchayati Raj institutions.
- It is responsible for local governance and development activities within a village or group of villages.
- The members of the Gram Panchayat are elected by the residents of the village through direct elections.

2. Panchayat Samiti or Mandal Parishad (Intermediate Level)

- The Panchayat Samiti (in some states called Mandal Parishad) functions at the intermediate level between the Gram Panchayat and the Zila Parishad.
- It oversees the administration and development activities of a group of villages within a block or mandal.
- Members of the Panchayat Samiti are elected by the members of the Gram Panchayats within its jurisdiction.

3. Zila Parishad (District Level)

- The Zila Parishad is the highest tier of the three-tier system of Panchayati Raj institutions.
- It is responsible for coordinating and supervising the functioning of the Gram Panchayats and Panchayat Samitis within the district.
- Members of the Zila Parishad are elected by elected members of the Panchayat Samitis and the members of the Legislative Assembly of the district.

The state of Rajasthan was the first to adopt the Panchayati Raj System on October 2, 1959. This was done through the Rajasthan Panchayat Raj Act, 1959. Rajasthan's adoption of Panchayati Raj marked a significant milestone in India's decentralization efforts. The state introduced a three-tier system of Panchayati Raj institutions, following the recommendations of the Balwant Rai Mehta Committee. Through this legislation, Gram Panchayats, Panchayat Samitis, and Zila Parishads were established, paving the way for local self-governance and rural development in the state.

Two tier system of Panchayat Raj

The transformation from a three-tier system of Panchayati Raj to a two-tier system occurred through amendments to the Panchayati Raj Acts in various states. This change typically involved the abolition of the intermediate tier, merging its functions with either the village-level or district-level Panchayats.

The rationale behind this transition varied from state to state, but some common reasons included administrative efficiency, reducing bureaucratic layers, and enhancing accountability. The transition from a three-tier to a two-tier system of Panchayati Raj occurred at different times in different states. It was a gradual process that took place over several years.

Madhya Pradesh (formerly known as the Central Provinces and Berar) was one of the first states to adopt the two-tier system of Panchayati Raj. The state introduced this change in 1962 through the Madhya Pradesh Panchayat Raj Act, abolishing the intermediate tier and implementing a two-tier system consisting of Gram Panchayats at the village level and Zila Parishads at the district level.

The transition to a two-tier system involved legislative amendments and administrative restructuring. In Madhya Pradesh, the state government passed the Madhya Pradesh Panchayat Raj Act, 1962, which abolished the Panchayat Samitis (intermediate level) and redistributed their functions to the Gram Panchayats and Zila Parishads.

Rationale behind transition

- *Administrative Efficiency:* Simplifying the structure by reducing the number of tiers was seen as a way to streamline decision-making and administrative processes.
- *Cost-effectiveness:* Eliminating the intermediate tier helped in reducing administrative costs associated with maintaining separate levels of Panchayati Raj institutions.
- *Accountability:* A two-tier system was believed to enhance accountability by establishing a direct link between local governance and district-level administration, thereby reducing potential confusion or overlapping of roles.

Overall, the transition from a three-tier to a two-tier system of Panchayati Raj was driven by a combination of administrative, financial, and governance considerations, with the aim of making local governance more efficient and responsive to the needs of rural communities.

14

Community Development

Community can be defined as a collective of individuals residing within a specific geographical area, sharing mutual interests and pursuits for sustenance. It represents a form of social organization that exists between the family unit and the broader state apparatus. While comprising various components within itself, a community also functions as an integral part of a larger societal framework. Dynamic in nature, a community is subject to alterations stemming from both internal dynamics and external influences. Several defining characteristics of a community include its cohesive nature, interrelated customs, intricate sub-group relationships, and the presence of discernible leadership structures.

Development, within the context of community, connotes a process of growth and maturation characterized by gradual and sequential phases of change. Understanding these foundational terms, community development programs can be conceptualized as initiatives aimed at facilitating gradual transformations within groups of individuals inhabiting a specific geographical region, united by shared livelihood pursuits and communal ties.

The notion of community development embodies a diverse range of interpretations, often encompassing endeavors directed towards the advancement of communal interests. As a compound term, it becomes imperative to dissect its constituent elements to gain comprehensive insights into its essence:

1. Community development serves as a concerted movement intended to enhance the overall quality of life for the entire community, propelled by active community involvement and initiative.
2. It encompasses a holistic program designed to stimulate local potential for growth across multiple dimensions, fostering reciprocal progress in both economic prosperity and social welfare. This progression hinges upon the intrinsic vitality of community members, with minimal reliance on external aid.

3. Technically facilitated yet locally driven, community development epitomizes a paradigm of self-help bolstered by external support systems.
4. On an international scale, the term encapsulates the collaborative process wherein grassroots efforts converge with governmental initiatives to ameliorate economic, social, and cultural conditions within communities, thereby integrating them into the national fabric and harnessing their contributions towards national progress.
5. It represents a strategic approach adopted by governments to engage rural populations effectively, harnessing local initiatives and energies to bolster productivity and elevate living standards.
6. Community development emerges as a dynamic process of social mobilization, wherein community members organize themselves to identify and address prevailing needs and challenges.

The inception of community development programs in India traces back to the post-independence era, with the initiation of the Community Development Program on October 2, 1952. Designed as a multi-faceted initiative aimed at fostering holistic rural development, this program encompassed diverse domains such as agriculture, animal husbandry, irrigation, cooperatives, public health, education, social education, communication, and village industries. The program's comprehensive scope encompassed facets of life pertinent to the majority of India's agrarian populace, with designated officials at the district level entrusted with the planning, execution, and evaluation of programmatic interventions down to the village level.

Key facets of the community development program underscore its overarching principles

1. It constitutes a nationwide endeavor targeting rural development.
2. A robust extension service, equipped with trained personnel, is strategically deployed at the village and block levels to facilitate and steer rural development initiatives.
3. Comprehensive schemes aimed at orienting and training personnel are instituted to equip the workforce with requisite competencies for effective program implementation.
4. The cultivation of local leadership through educational initiatives assumes paramount importance, aligning with the ethos of community development.

5. The integration of youth programs within the community development framework emerges as a pivotal strategy, fostering rural youth's holistic development and empowerment.

Central to the community development paradigm lies its commitment to fostering grassroots participation and empowerment, guided by a set of core principles elucidated by Michael and Julie Bopp:

1. The adept management of community tensions serves as a catalyst for constructive change.
2. Facilitative processes that engender open consultation regarding community realities and needs are imperative for informed decision-making.
3. The preservation of unity and healthy interpersonal relations forms the bedrock of sustainable community development endeavors.
4. A shared vision of a sustainable future serves as a guiding beacon, driving collective efforts towards common goals.
5. The cultivation of core groups within communities plays a pivotal role in galvanizing local initiatives and effecting transformative change.
6. Personal revitalization and healing are integral components of community development endeavors, fostering individual agency and resilience.
7. Facilitating learning experiences within the community fosters knowledge dissemination and capacity building.
8. The establishment of effective organizational structures enhances the efficiency and efficacy of community-led initiatives.
9. Networking with external resources and allies amplifies the impact and reach of community development efforts.

Programmatic development within the community development framework is complemented by a reflective process encompassing monitoring and evaluation, safeguarding the integrity of the programmatic trajectory.

The philosophy underpinning community development programs is grounded in several foundational tenets:

1. Initiatives are rooted in addressing felt needs, ensuring programs resonate with community aspirations.
2. The innate desire for emancipation from poverty and suffering serves as a driving force, underpinning community-driven initiatives.

3. Fundamental freedoms, including autonomy in shaping economic, religious, educational, and political institutions, are upheld as integral to community development endeavors.
4. People's values are accorded due reverence, shaping programmatic interventions to align with local ethos and aspirations.
5. Self-help emerges as a foundational principle, underscoring communities as reservoirs of resilience and ingenuity.
6. Human capital, represented by the collective agency of community members, is deemed the most valuable resource, driving sustainable development efforts.
7. Community development encompasses a transformative process, catalyzing shifts in attitudes, behaviors, and knowledge domains essential for holistic advancement.

The evolutionary trajectory of community development programs unfolds across three discernible stages:

1. *Relationship Building:* Establishing trust, cohesion, and shared objectives within the community forms the foundational stage.
2. *Process Development:* Engaging in participatory processes to identify needs, formulate strategies, and effect change characterizes this phase.
3. *The Consolidation of Structures and Mechanisms:* Institutionalizing sustainable frameworks and mechanisms to perpetuate community-led initiatives marks the culmination of the developmental trajectory.

Despite its myriad merits, community development programs encounter several constraints:

1. Widespread illiteracy poses a formidable challenge, necessitating comprehensive training initiatives to capacitate rural populations.
2. Infrastructural deficiencies, including inadequate communication channels and transportation networks, impede programmatic reach and efficacy.
3. Resource constraints, both in terms of funding and trained personnel, pose significant hurdles to program implementation and sustainability.
4. Prevailing societal norms and entrenched practices often inhibit the embrace of transformative initiatives, necessitating nuanced engagement strategies.

5. The dearth of authentic leadership within traditional, agrarian societies poses a formidable impediment to effective programmatic implementation.

6. The dissemination of new practices and behavioral norms mandates judicious communication strategies, underscored by an acute understanding of change dynamics and community psychology.

In conclusion, community development programs represent a holistic approach to grassroots empowerment and sustainable development, underpinned by principles of participatory governance, self-reliance, and social equity. Grounded in collaborative endeavors and guided by a profound understanding of community dynamics, these initiatives hold immense potential for fostering inclusive growth and resilience within diverse societal contexts.

15

Rural Leadership

Meaning of Leader

Leadership is a multifaceted concept that involves influencing, guiding, and inspiring individuals or groups towards a common goal or vision. It encompasses a range of qualities and behaviors that enable a person to effectively mobilize and motivate others.

Characteristics of a Leader

Leadership characteristics can vary based on individual styles, contexts, and goals, but some key traits commonly associated with effective leaders include:

1. *Visionary:* Leaders have a clear vision of what they want to achieve and can communicate it effectively to others.
2. *Integrity:* Leaders act with honesty, fairness, and transparency, earning the trust and respect of their team members.
3. *Empathy:* Understanding the perspectives, needs, and concerns of others allows leaders to build strong relationships and foster collaboration.
4. *Communication skills:* Effective leaders are adept communicators who can convey ideas, provide feedback, and listen actively to others.
5. *Confidence:* Confidence in oneself and in the team's abilities inspires trust and motivation among team members.
6. *Decisiveness:* Leaders make timely and well-informed decisions, even in the face of uncertainty or ambiguity.
7. *Resilience*: Leaders demonstrate resilience in the face of challenges, setbacks, or failures, bouncing back and guiding their team through adversity.
8. *Adaptability*: Leaders are flexible and open to change, able to adjust their strategies and approaches as needed to achieve goals.

9. *Empowerment*: Effective leaders empower their team members, delegating authority and providing support to help them grow and succeed.
10. *Accountability*: Leaders take responsibility for their actions and decisions, holding themselves and others accountable for achieving results.
11. *Courage*: Courageous leaders are willing to take risks, challenge the status quo, and advocate for what they believe is right, even in the face of opposition.
12. *Influence*: Leaders have the ability to influence others' attitudes, behaviors, and actions towards achieving shared objectives.

These characteristics can manifest in various ways depending on the leadership style and the specific context in which a leader operates. However, they are often foundational to effective leadership across different domains.

Rural Leadership

The concept of rural leadership refers to the unique set of skills, qualities, and actions required to effectively lead within rural communities or settings. Rural leadership encompasses the following key aspects:

1. *Understanding of Rural Context:* Effective rural leaders have a deep understanding of the challenges, opportunities, and dynamics specific to rural areas. They recognize the importance of factors such as agriculture, natural resources, small businesses, and community cohesion in rural life.
2. *Community Engagement:* Rural leadership often involves close engagement with the local community. Leaders build relationships, foster trust, and collaborate with community members, organizations, and stakeholders to address shared concerns and work towards common goals.
3. *Advocacy for Rural Issues:* Leaders in rural areas advocate for the needs and interests of their communities at local, regional, and sometimes national levels. This may involve lobbying for infrastructure improvements, access to healthcare and education, economic development initiatives, and other policies that support rural prosperity.
4. *Innovation and Problem-Solving:* Effective rural leaders are innovative and resourceful problem-solvers. They seek creative solutions to

address the unique challenges facing rural communities, such as limited access to services, declining populations, economic disparities, and environmental concerns.

5. *Empowerment and Capacity Building:* Rural leaders empower community members by providing opportunities for skill development, leadership training, and civic engagement. They encourage local residents to take ownership of community projects and initiatives, fostering a sense of pride and self-reliance.
6. *Collaboration and Partnerships:* Leadership in rural areas often requires collaboration and partnerships with diverse stakeholders, including government agencies, nonprofit organizations, businesses, educational institutions, and neighboring communities. Leaders facilitate dialogue, build consensus, and leverage resources to achieve collective goals.
7. *Cultural Sensitivity and Inclusivity:* Rural leaders respect and value the diversity of perspectives, cultures, and traditions within their communities. They strive to ensure that all voices are heard and represented in decision-making processes, promoting inclusivity and social cohesion.
8. *Resilience and Adaptability:* Rural leadership demands resilience and adaptability in the face of challenges such as economic downturns, natural disasters, demographic shifts, and changing social dynamics. Leaders remain flexible, proactive, and resilient, seeking opportunities for growth and development even in uncertain times.

Effective rural leadership is essential for promoting sustainable development, enhancing quality of life, and preserving the unique character and vitality of rural communities. It requires a combination of empathy, vision, collaboration, and strategic thinking to address the complex and interconnected issues facing rural areas.

Functions of Leaders

The functions of a leader encompass a range of responsibilities and actions that are essential for effective leadership. These functions can vary depending on the specific context, goals, and organizational structure, but they generally include the following:

1. *Setting Direction and Vision:* Leaders establish a clear direction and vision for the organization, team, or community. They articulate goals, priorities, and strategies to guide decision-making and action.

2. *Motivating and Inspiring:* Leaders inspire and motivate individuals or groups to achieve their full potential and pursue common objectives. They foster enthusiasm, commitment, and a sense of purpose among team members.
3. *Providing Guidance and Support:* Leaders offer guidance, support, and mentorship to help individuals develop their skills, overcome challenges, and achieve personal and professional growth.
4. *Facilitating Communication:* Leaders facilitate open, honest, and effective communication within the organization or team. They encourage dialogue, exchange of ideas, and feedback to promote understanding and collaboration.
5. *Making Decisions:* Leaders make timely and well-informed decisions, often in complex or ambiguous situations. They gather relevant information, weigh options, and consider the potential impact on stakeholders before taking action.
6. *Problem-Solving:* Leaders identify and address problems, obstacles, and conflicts that arise within the organization or team. They employ analytical thinking, creativity, and collaboration to find viable solutions.
7. *Building Relationships and Networks:* Leaders cultivate positive relationships and networks with internal and external stakeholders, including team members, clients, partners, and community members. They nurture trust, collaboration, and mutual respect to enhance organizational effectiveness and achieve common goals.
8. *Empowering Others:* Leaders empower individuals or teams by delegating authority, fostering autonomy, and providing resources and support. They create an environment that encourages initiative, innovation, and accountability.
9. *Managing Change:* Leaders anticipate and manage change effectively, whether it involves internal restructuring, technological advancements, market shifts, or other external factors. They communicate the rationale for change, address concerns, and facilitate adaptation and resilience among team members.
10. *Promoting Diversity and Inclusion:* Leaders champion diversity and inclusion within the organization or team, valuing and respecting the unique perspectives, backgrounds, and experiences of all members. They create an inclusive culture that celebrates diversity and fosters belonging.

11. *Modeling Behavior:* Leaders lead by example, demonstrating the values, ethics, and behaviors they expect from others. They embody integrity, professionalism, and a commitment to excellence in their actions and decisions.

12. *Evaluating Performance and Providing Feedback:* Leaders assess individual and team performance against established goals and expectations. They provide constructive feedback, recognition, and support to help individuals improve and excel in their roles.

The functions of a leader are multifaceted and interconnected, requiring a combination of strategic thinking, interpersonal skills, and emotional intelligence to effectively guide and inspire others towards shared objectives.

Classification of Leaders

Certainly! Leadership can be classified in various ways based on different criteria. Let's explore some common classifications of leaders:

1. By Leadership Style

- *Autocratic Leaders:* These leaders make decisions independently without consulting others. They typically have a high degree of control over their team and may not welcome input from others.
- *Democratic Leaders*: Democratic leaders involve their team members in the decision-making process, seeking their input and opinions before making a final decision.
- *Laissez-faire Leaders*: Laissez-faire leaders provide minimal guidance or direction to their team members, allowing them to make their own decisions and manage their own tasks autonomously.

2. By Organizational Hierarchy

- *Top-Level Leaders:* These leaders occupy senior positions within an organization, such as CEOs, presidents, or executive directors. They are responsible for setting the overall direction and strategy of the organization.
- *Middle-Level Leaders:* Middle-level leaders, such as department managers or team leaders, bridge the gap between top-level leadership and front-line employees. They implement strategies, coordinate activities, and manage day-to-day operations within their respective departments or teams.

- *Front-Line Leaders:* Front-line leaders directly supervise and oversee the work of individual employees or small teams. They are responsible for guiding and supporting their team members to achieve specific goals and objectives.

3. By Leadership Approach

- *Transactional Leaders:* Transactional leaders focus on exchange relationships with their followers, offering rewards or incentives in exchange for compliance with instructions or performance of tasks.
- *Transformational Leaders:* Transformational leaders inspire and motivate their followers by articulating a compelling vision, challenging the status quo, and fostering innovation and growth.
- *Servant Leaders:* Servant leaders prioritize the needs and well-being of their followers, serving as mentors, coaches, and facilitators to support their personal and professional development.

4. By Leadership Effectiveness

- *Effective Leaders:* Effective leaders achieve desired outcomes and objectives, such as high team performance, productivity, and satisfaction.
- *Ineffective Leaders:* Ineffective leaders fail to achieve desired results or may even have a negative impact on their team or organization due to poor decision-making, communication, or interpersonal skills.

5. By Leadership Focus

- *Task-Oriented Leaders:* Task-oriented leaders prioritize the accomplishment of specific goals, objectives, or tasks within the organization. They focus on planning, organizing, and directing activities to achieve desired outcomes.
- *People-Oriented Leaders:* People-oriented leaders prioritize the well-being, development, and satisfaction of their team members. They focus on building positive relationships, fostering collaboration, and supporting the personal and professional growth of individuals.

Leadership is a complex phenomenon, and individuals may exhibit traits or behaviors from multiple classifications depending on the situation and context. Understanding different classifications of leaders can help us recognize and appreciate the diversity of leadership styles and approaches in organizations and communities. Different authors classified the leaders in different ways. Some of the classificationsare:

(A) ***Beal, Bohlen and Raudabaugh (1962)*** classified the leaders into four types as;

a) Born leaders

b) Passive leaders (personal power or characteristic leader)

c) Bureaucratic leaders

d) Democratic leaders

(B) ***Mott (1972)*** spoke about three major leadership categories

a) Democratic

b) Multifactor

c) Situational

(C) ***Henning (1962)*** analyzed the leadership and given the following three:

a) Autocrat

b) Bureaucrat

c) Neurocrat

(D) ***Lester (1975)*** also pointed three types as:

a) Autocratic

b) Democratic

c) Free rein

(E) ***Haiman (1951)*** described five categories as the

a) Executive

b) Judge

c) Advocate

d) Expert

e) Discussion leader

(F) **Carter (1961)** identified three leadership patterns according to the manner inwhich the leader was mostly oriented. They are,

a) Personal

b) Institutional

c) Flexible

(G) ***Sachdeva and Vidyabhushan (1974)*** studied three main leadership types as;

a) Authoritarian

b) Democratic

c) Laissez-faire

(H) ***Dahama and Bhatnagar (1985)*** found several ways of classifying leaders, someof them are;

a) Democratic,

b) Autocratic

c) Laissez-faire

d) Formal and informal leaders

e) Professional

f) Lay (voluntary) leaders

g) Political,

h) Religious,

i) Social

j) Academic

k) Elected,

l) Selected or nominated

m) Popular

n) Unpopular

o) Traditional

p) Progressive leaders

16

Extension Administration

Extension administration refers to the management and coordination of extension services provided to communities, typically in agricultural, rural development, and related fields. Extension services aim to disseminate knowledge, information, and technology to help improve the livelihoods, productivity, and well-being of individuals and communities.

Extension Administration Involves Several Key Functions

1. *Program Development:* Extension administrators are responsible for identifying the needs and priorities of target communities and developing programs and initiatives to address them. This involves conducting needs assessments, setting goals and objectives, and designing interventions or services tailored to local contexts.
2. *Resource Allocation:* Extension administrators allocate human, financial, and material resources to support the implementation of extension programs. They manage budgets, procure supplies and equipment, and allocate staff to ensure the effective delivery of services.
3. *Staff Management:* Extension administrators recruit, train, supervise, and evaluate extension personnel, including extension agents, specialists, and support staff. They provide guidance, mentorship, and professional development opportunities to ensure that staff are equipped to deliver high-quality services.
4. *Coordination and Collaboration:* Extension administrators collaborate with various stakeholders, including government agencies, non-governmental organizations, research institutions, and community groups, to leverage resources, expertise, and networks in support of extension programs. They coordinate activities, share information, and build partnerships to maximize the impact of extension efforts.
5. *Monitoring and Evaluation:* Extension administrators monitor and evaluate the performance and impact of extension programs to assess their effectiveness, identify strengths and weaknesses, and make informed

decisions for improvement. This involves collecting data, analyzing results, and using feedback to inform programmatic adjustments and strategic planning.

6. *Communication and Outreach:* Extension administrators communicate with target audiences and stakeholders to raise awareness of extension services, disseminate information, and promote participation and engagement. They use various channels, including workshops, field days, publications, and digital media, to reach diverse audiences and facilitate knowledge exchange.
7. *Policy Advocacy:* Extension administrators advocate for policies and regulations that support the goals and objectives of extension programs. They engage with policymakers, legislators, and other decision-makers to influence policies related to agriculture, rural development, environmental conservation, and other relevant areas.

Overall, extension administration plays a crucial role in facilitating the delivery of extension services and promoting sustainable development and empowerment in rural communities. Effective administration ensures that extension programs are well-designed, efficiently managed, and responsive to the evolving needs of target populations.

Difference Between Management & Administration

Extension management and extension administration are closely related concepts, but they focus on different aspects of the extension service delivery process. Here's a breakdown of the key differences between the two:

1. Focus

- *Extension Management:* Extension management primarily focuses on the operational aspects of running extension programs and services. It involves activities such as planning, organizing, staffing, directing, and controlling resources to ensure the effective implementation of extension activities.
- *Extension Administration:* Extension administration, on the other hand, has a broader focus that encompasses both managerial and strategic functions related to extension services. It involves tasks such as program development, policy formulation, resource allocation, coordination, monitoring, evaluation, and advocacy.

2. Scope

- *Extension Management:* The scope of extension management is more operational and tactical. It involves day-to-day decision-making and activities related to the implementation of extension programs, as well as the supervision and management of personnel and resources at the field level.
- *Extension Administration:* Extension administration has a broader scope that includes strategic planning and decision-making at the organizational or institutional level. It involves setting goals and priorities, allocating resources, establishing policies and procedures, and ensuring compliance with regulatory requirements.

3. Responsibilities

- *Extension Management:* The responsibilities of extension management typically include managing field staff, organizing training programs and workshops, coordinating outreach activities, liaising with stakeholders, and overseeing the delivery of extension services in specific geographic areas or communities.
- *Extension Administration:* Extension administration encompasses a wider range of responsibilities, including setting overall strategic direction, developing long-term plans and budgets, securing funding, establishing partnerships, advocating for policy changes, and ensuring accountability and transparency in program implementation.

4. Decision-Making Authority

- *Extension Management*: Decision-making authority in extension management is often decentralized, with field-level managers and supervisors making decisions related to day-to-day operations and resource allocation within their areas of responsibility.
- *Extension Administration*: Decision-making authority in extension administration is typically centralized at higher levels of the organization or institution, where administrators and executives make strategic decisions related to program priorities, resource allocation, policy formulation, and organizational development.

In general, extension management deals with the operational aspects of implementing extension programs at the field level, while extension administration focuses on strategic planning, policy formulation, and organizational management at higher levels. Both functions are essential for the effective delivery of extension services and the achievement of desired outcomes in agricultural, rural development, and related fields.

S. No	Basis	Administration	Management
1.	Meaning	Administration is concerned with the formulation ofobjectives, plans the work and policies of the organization	Management means getting done through and with others.
2.	Nature of work	Administration relates to the decision- making. It is a thinking function.	Management refers to execution of decisions. It is adoing function.
3.	Decision-making	Administration determines what is to be done and whenit is to be done	Management decides who shall implement the administrativedecisions.
4.	Status	Administration refers tohigher levels of management	Management is relevant at lower levels in the organization.

Principles of Management

Henri Fayol, often hailed as the father of modern management, laid the foundation for a body of management principles based on his extensive practical experience as a manager. While he did not formulate a unified theory of management, his principles resonate remarkably with contemporary management theories. Fayol proposed the existence of a singular "administrative science" whose principles could be universally applied across diverse management contexts, irrespective of the type of organization under consideration. This concept earned him the epithet of "Universality." Nevertheless, Fayol underscored that his principles were not immutable laws but rather pragmatic guidelines to be flexibly employed as circumstances dictated.

Fayol identified six broad categories into which the activities of an industrial enterprise could be classified:

(i) Technical (pertaining to production)

(ii) Commercial (encompassing buying, selling, and exchange)

(iii) Financial (pertaining to the acquisition and optimal utilization of capital)

(iv) Security (ensuring the protection of property and individuals)

(v) Accounting (inclusive of statistical analysis)

(vi) Managerial.

However, his primary focus rested on managerial activities, and he formulated a set of overarching principles applicable to all organizational domains:

(i) Authority and Responsibility Nexus

(ii) Unity of Command

(iii) Unity of Direction

(iv) Scalar Chain of Command

(v) Division of Labor

(vi) Discipline

(vii) Subordination of Individual Interests to the Collective Good

(viii) Remuneration

(ix) Centralization

(x) Orderliness

(xi) Equity

(xii) Stability of Tenure of Personnel

(xiii) Initiative

(xiv) Esprit de Corps

These principles elucidate the fundamental tenets underpinning effective management across diverse organizational settings, emphasizing the importance of clear lines of authority, coordinated action, efficient division of labor, and fostering a conducive work environment characterized by mutual respect and collective endeavor. Fayol's enduring legacy lies in his profound insights into the intricacies of organizational management, which continue to guide managerial practices to this day.

Management Functions & Process of Management

Henri Fayol delineates five primary functions of management: planning, organizing, commanding, coordinating, and controlling. Luther Gulick expands upon this framework, identifying seven functions encapsulated in the acronym "POSDCORB": planning, organizing, staffing, directing, coordinating, reporting, and budgeting. For our analytical purposes, we will focus on six core functions of a manager: planning, organizing, staffing, directing, coordinating, and controlling.

Planning stands as the foundational and overarching function among all management activities. It entails delineating future objectives, strategizing courses of action, and establishing timelines for implementation. The planning process encompasses the identification of both long-term and short-term goals, the development of actionable strategies to achieve said objectives, and the formulation of policies and procedures to guide implementation.

Organizing involves the systematic arrangement of activities required to achieve organizational objectives. This includes grouping activities into coherent jobs, assigning responsibilities and authority to individuals and departments, and fostering coordination among various organizational units. Key sub-functions within organizing encompass activity identification, job grouping, employee assignment, delegation of authority, and establishment of coordination mechanisms.

Staffing emerges as a continuous and critical function essential for securing qualified personnel to fulfill organizational roles effectively. This function encompasses various sub-components such as manpower planning, recruitment, selection, placement, training, and career development. Staffing plays a pivotal role in ensuring organizational efficiency and effectiveness by aligning personnel capabilities with job requirements.

Directing entails the guidance and motivation of employees to perform optimally in pursuit of organizational goals. This function involves clarifying job expectations, providing guidance on job performance, and fostering employee motivation and engagement. Sub-functions within directing include communication, motivation, and leadership, all aimed at eliciting optimal employee performance.

Coordination represents the harmonization of organizational activities to ensure alignment with overarching objectives. This function entails establishing relationships among organizational units to facilitate cohesive action towards common goals. Sub-functions within coordination include defining authority-responsibility relationships, fostering unity of direction and command, facilitating effective communication, and providing leadership.

Controlling serves as the mechanism for ensuring organizational performance remains aligned with predetermined objectives and goals. This function encompasses the measurement of performance against established benchmarks, identification of deviations, and implementation of corrective actions. Through effective control mechanisms, organizations can maintain operational efficiency and adapt to changing circumstances.

Reporting mandates the timely dissemination of relevant information to stakeholders within and outside the organization. Executives are responsible for keeping stakeholders informed of organizational activities and performance through records, reports, and inspections.

Budgeting represents a critical aspect of organizational management, facilitating fiscal planning, accounting, and financial control. It involves the allocation of resources in alignment with organizational priorities, ensuring optimal utilization and accountability.

In summary, the functions of a manager encompass a comprehensive range of activities aimed at achieving organizational objectives. By effectively planning, organizing, staffing, directing, coordinating, controlling, reporting, and budgeting, managers can steer their organizations towards sustained success and competitive advantage.

17

Monitoring and Evaluation

Monitoring

Monitoring in agricultural extension refers to the systematic and ongoing process of collecting, analyzing, and evaluating data and information to assess the progress, effectiveness, and impact of extension programs and activities. It involves tracking key indicators, outputs, and outcomes to ensure that extension services are being delivered efficiently and are contributing to the desired outcomes for farmers and rural communities.

Several famous authors have provided definitions of monitoring in the context of agricultural extension:

1. *Rogers and Shoemaker (1971):* "Monitoring refers to the systematic collection, analysis, and interpretation of data on the degree to which extension goals are being achieved."
2. *World Bank (1992):* "Monitoring involves the routine tracking of outputs and progress toward objectives to detect potential problems and suggest modifications."
3. *Adekoya (2002):* "Monitoring is a process of systematically collecting, analyzing and using information for management and decision-making. It is the continuous collection of data aimed at providing early indications of the need for corrective action."

Advantages of monitoring in agricultural extension include:

4. *Performance Evaluation:* Monitoring allows extension organizations to assess the performance of extension programs, activities, and personnel against predetermined goals, targets, and indicators. This helps identify strengths, weaknesses, and areas for improvement.
5. *Accountability:* Monitoring provides evidence-based information on the utilization of resources and the achievement of results, enhancing transparency and accountability in extension service delivery. It allows extension organizations to demonstrate accountability to funders, policymakers, and other stakeholders.

6. *Decision-Making:* Monitoring data and information enable extension managers and decision-makers to make informed decisions about program design, resource allocation, and strategic planning. It helps identify emerging issues, trends, and priorities that require attention or intervention.
7. *Quality Improvement:* By continuously monitoring program implementation and outcomes, extension organizations can identify bottlenecks, gaps, or inefficiencies in service delivery and take corrective actions to improve the quality and effectiveness of extension services.
8. *Learning and Innovation:* Monitoring fosters a culture of learning, reflection, and innovation within extension organizations. It provides opportunities to capture and share lessons learned, best practices, and success stories, facilitating organizational learning and adaptation.
9. *Stakeholder Engagement:* Monitoring promotes stakeholder engagement and participation by involving farmers, communities, policymakers, and other stakeholders in the monitoring process. It fosters collaboration, dialogue, and shared responsibility for achieving development goals.

Monitoring plays a critical role in ensuring the accountability, effectiveness, and sustainability of agricultural extension programs, ultimately contributing to improved livelihoods, food security, and rural development.

Evaluation

Evaluation in agricultural extension refers to the systematic assessment of the effectiveness, efficiency, relevance, and sustainability of extension programs, activities, and interventions. It involves collecting, analyzing, and interpreting data and information to determine the extent to which extension objectives are being achieved and to inform decision-making and program improvement.

The term "evaluation" has its origins in the Latin word "evaluare," which means "to assess" or "to determine the value of." Evaluation as a concept has been used across various disciplines and fields for centuries, but its formal application in the context of agricultural extension began to gain prominence in the 20th century.

Several famous authors have provided definitions of evaluation in the context of agricultural extension:

1. Rogers and Shoemaker (1971): "Evaluation refers to the systematic application of social research procedures to assess the conceptualization, design, implementation, and utility of communication campaigns, often

with emphasis on the structure, content, and outcomes of the campaign messages."

2. World Bank (1992): "Evaluation involves the systematic and objective assessment of an ongoing or completed project, program, or policy, its design, implementation, and results. The aim is to determine the relevance and fulfillment of objectives, developmental efficiency, effectiveness, impact, and sustainability."
3. Adekoya (2002): "Evaluation is the systematic assessment of the design, implementation, and outcomes of a project, program, or policy. It involves gathering and analyzing information on the extent to which the project, program, or policy has achieved its intended objectives and contributed to desired outcomes."

These definitions highlight the key elements of evaluation, including its systematic nature, its focus on assessing various aspects of program design and implementation, and its emphasis on determining the extent to which objectives are being met and desired outcomes are being achieved.

Evaluation serves several purposes in agricultural extension, including

1. *Assessing Impact:* Evaluation helps determine the extent to which extension interventions are achieving their intended outcomes and contributing to desired changes in farmer knowledge, attitudes, practices, and livelihoods.
2. *Informing Decision-Making:* Evaluation provides evidence-based information to extension managers, policymakers, and other stakeholders to inform decision-making, resource allocation, and program improvement efforts.
3. *Accountability and Transparency:* Evaluation promotes accountability and transparency by assessing the performance and results of extension programs and activities and communicating findings to stakeholders.
4. *Learning and Improvement:* Evaluation facilitates organizational learning and improvement by identifying lessons learned, best practices, and areas for improvement in program design, implementation, and management.

Evaluation ensures the effectiveness, relevance, and sustainability of agricultural extension programs, ultimately contributing to improved agricultural productivity, food security, and rural development.

Types of Evaluation

Process Evaluation: Process evaluation serves to gauge the activities of a program, assess program quality, and determine its target audience. As outlined by Hawe and colleagues, this evaluation method aids in addressing pivotal questions regarding the program's efficacy:

- Has the project effectively reached its target group?
- Are all project activities reaching every segment of the target group?
- Are participants and key stakeholders satisfied with all aspects of the project?
- Are all activities being executed as intended, and if not, what are the reasons?
- What modifications, if any, have been made to the intended activities?
- Are all materials, information, and presentations suitable for the target audience?

Impact Evaluation: Impact evaluation focuses on measuring the immediate effects of a program and aligns with its objectives. It assesses the extent to which the program objectives, including sub-objectives, have been realized. This evaluation method addresses questions such as:

- How effectively has the project achieved its objectives, including sub-objectives?
- To what extent have the desired short-term changes been accomplished?

Impact evaluation entails measuring the program's effectiveness immediately after its completion and up to six months thereafter.

Outcome Evaluation: Outcome evaluation delves into the long-term effects of the program and typically evaluates the program's overarching goal. It assesses the degree to which the program goal has been accomplished, considering factors outside the program that may have influenced or impeded the desired change. Key questions addressed by outcome evaluation include:

- Has the overall program goal been attained?
- What external factors, if any, have contributed to or hindered the desired change?
- What unintended consequences, if any, have arisen from the program?

Outcome evaluation evaluates changes occurring at least six months after program implementation, focusing on longer-term outcomes. While primarily assessing the main goal, outcome evaluation can also track program objectives over time.

Summative Evaluation: Summative evaluation occurs at the conclusion of a program cycle and provides a comprehensive overview of program effectiveness. It examines program outcomes to determine overall effectiveness and informs decision-making regarding program continuation or modification. Summative evaluation addresses questions such as:

- Have the program objectives been met?
- What modifications are required to enhance program structure?
- What is the overall impact of the program?
- What resources are needed to address program weaknesses?
- Is program continuation warranted, and if so, in what capacity?
- Can the program be replicated in other settings?
- How sustainable is the program?
- What factors have influenced program outcomes?
- What recommendations emerge from the program?

Summative evaluation informs decisions concerning specific services and the future trajectory of the program, which may not be feasible during the program cycle's midpoint. Summative evaluations are disseminated to funders and stakeholders with a vested interest in the program.

Formative Evaluation: Formative evaluation is an ongoing process that enables real-time feedback implementation during a program cycle. It focuses on examining and adapting processes as they unfold, providing timely insights to fine-tune program strategies. Formative evaluations facilitate program adjustments "on the fly" to aid in achieving program goals.

Indicators for Evaluation

Indicators are essential for determining the data necessary to assess program progress and alignment with goals and objectives. Evaluation indicators fall into two categories:

Process Indicators: Process indicators monitor program implementation and inputs, including financial, human, and administrative resources, and required equipment. These indicators assess how well the program is implemented, reaching the intended target, and maintaining acceptable quality. Process indicators encompass:

- Number of participants
- Proportion of the target population participating
- Attendance rates
- Dropout rates
- Involvement of key stakeholders
- Participant satisfaction, assessed by:
- Comfort, understanding, and receptiveness of participants
- Friendliness and approachability of participants and staff
- Appropriateness of venue and setup
- Program affordability and scheduling convenience
- Relevance and interest of program topics
- Program implementation, assessed by:
- Number of workshops conducted
- Implementation of all activities
- Effectiveness and suitability of materials
- Media coverage achieved

Impact and Outcome Indicators: Impact indicators monitor progress toward program objectives, often focusing on short-term changes in knowledge, attitudes, and intended behavior. Outcome indicators assess the achievement of program goals, including actual behaviors, health status, and quality of life, reflecting longer-term changes or sustained changes over time. Impact indicators may include:

- Awareness, knowledge, and skills changes
- Intended behavior changes
- Individual capacity improvements, such as confidence and problem-solving skills
- Enhanced social networks and relationships

- Outcome indicators may encompass:
- Improved mental and physical well-being
- Increased community engagement, education, and employment

Purposes of Evaluation

1) Ensuring planned results are achieved
2) Monitoring progress toward objectives
3) Improving and supporting management
4) Building shared understanding
5) Generating new knowledge and supporting learning
6) Capacity building
7) Motivating stakeholders
8) Ensuring accountability
9) Garnering public and political support
10) Guiding management decision-making
11) Promoting organizational learning
12) Soliciting support for programs

Challenges of Evaluation

1) Assessing long-term impacts
2) Addressing uncertainty
3) Harmonizing diverse agendas
4) Simplifying complexity
5) Cultivating a learning culture
6) Navigating political imperatives
7) Overcoming capacity constraints
8) Managing conflicts

Differences between Monitoring and Evaluation

Monitoring	Evaluation
Monitoring is the systematic and routine collection of information about the programs/projects activities	Evaluation is the periodic assessment of the programs/projects activities
It is ongoing process which is done to see if things/activities are going on track or not i.e. it regularly tracks the program	It is done on a periodic basis to measure the success against the objective i.e. it is an in-depth assessment of the program
Monitoring is to be done starting from the initial stage of the projects	Evaluation is to be done after certain point of time of the project, usually at the mid of the project, completion of the project or while moving from one stage to another stage of the projects/programs
Monitoring is done usually by the internal members of the team	Evaluation is done mainly done by theexternal members. However, sometimes it may be also done by internal members of theteam or by both internal and external members in a combined way
Monitoring provides information about the current status and thus helps to take immediate remedial actions, if necessary	Evaluation provides recommendations, information for long term planning and lessons for organizational growth and success
It focuses on input, activities and output	It focuses on outcomes, impacts and overall goal
Monitoring process includes regular meetings, interview, monthly and quarterly reviews etc. Usually quantitativedata.	Evaluation process includes intense data collection, both qualitative and quantitative
It has multiple points of data collection	Data collection is done at intervals only
It gives answer about the present scenario of the project towards achieving planned results considering the human resources, budget, materials, activities and outputs	It assesses the relevance, impact, sustainability, effectiveness and efficiency of the projects
Monitoring studies the presentinformation and experiences of the project	Evaluation studies the past experience of the project performance
Monitoring checks whether the project did what it said it would do	Evaluation checks whether what the project did had the impact that it intended
Helps to improve project design and functioning of current project	Helps to improve project design of future projects
Monitoring looks at detail of activities	Evaluation does not look at detail of activities but rather looks at a bigger picture
It compares the current progress with the planned progress	It looks at the achievement of the programs along with both positive/negative, intended/ unintended effects
Information obtained from monitoring is more useful to theimplementation/ management team	Information obtained from evaluation isuseful to all the stakeholders

Monitoring	Evaluation
Monitoring result is used for informed actions and decisions	Evaluation result is used for planning of new programs and interventions
Answers the question "Are we doing things right?"	Answers the question "Are we doing right thing?"
Regular report and updates about the project/program act a deliverables here	Reports with recommendations and lessonsact as a deliverable here
Good or effective monitoring does notrely on evaluation results	Good or effective evaluation relies to some extent on good monitoring
There are few quality checks inmonitoring	There are many quality checks in evaluation
It provides information for evaluation	It provides information for proper planning

Methods and tools of Monitoring & Evaluation

Methods and tools for monitoring and evaluation (M&E) in agricultural extension play a crucial role in assessing the effectiveness, efficiency, relevance, and impact of extension programs and activities. These methods and tools help extension organizations collect, analyze, and interpret data and information to inform decision-making, program improvement, and accountability. Here's a brief overview of some commonly used methods and tools for M&E in agricultural extension:

1. *Surveys and Questionnaires*: Surveys and questionnaires are widely used tools for collecting data from farmers and other stakeholders. They can be administered through face-to-face interviews, phone calls, or online platforms to gather information on farmer demographics, knowledge, attitudes, practices, and satisfaction with extension services.

2. *Interviews and Focus Group Discussions (FGDs)*: Interviews and FGDs involve direct interaction with farmers and stakeholders to gather qualitative data on their experiences, perceptions, and feedback regarding extension programs and interventions. These methods provide valuable insights into the context, challenges, and opportunities for agricultural development.

3. *Field Visits and Observations*: Field visits and observations allow extension personnel to directly observe and assess the implementation of extension activities, such as training workshops, field demonstrations, and farmer meetings. They provide firsthand insights into program delivery, participant engagement, and the effectiveness of extension methods and approaches.

4. *Participant and Beneficiary Registers*: Participant and beneficiary registers are used to keep track of individuals and groups who participate

in extension programs and activities. They help monitor program reach, attendance rates, and demographic information, facilitating targeted interventions and follow-up support.

5. *Monitoring Checklists and Tools*: Monitoring checklists and tools provide standardized criteria and indicators for assessing the quality, coverage, and progress of extension activities. They help ensure consistency and uniformity in data collection and enable systematic tracking of key performance indicators.
6. *Performance Indicators and Targets*: Performance indicators and targets are quantitative measures used to assess progress and achievements against predefined objectives and benchmarks. They help monitor program outputs, outcomes, and impact over time and provide a basis for comparison and analysis.
7. *Data Management Systems*: Data management systems, including databases, spreadsheets, and software applications, are used to organize, store, and analyze monitoring and evaluation data. They facilitate data entry, storage, retrieval, and analysis, ensuring data quality, accuracy, and security.
8. *Participatory M&E Approaches*: Participatory M&E approaches involve engaging farmers and stakeholders in the monitoring and evaluation process. They promote ownership, transparency, and accountability by involving beneficiaries in data collection, analysis, interpretation, and decision-making.
9. *Remote Sensing and Geospatial Analysis*: Remote sensing and geospatial analysis technologies, such as satellite imagery, geographic information systems (GIS), and global positioning system (GPS) data, are used to assess land use, crop health, and environmental changes. They provide spatially explicit information for M&E activities and support evidence-based decision-making in agricultural development.

By utilizing these methods and tools effectively, extension organizations can strengthen their M&E systems, enhance program performance, and maximize the impact of extension services on farmer livelihoods, food security, and rural development.

18

Transfer of Technology

Transfer of Technology

Technology transfer encompasses the process by which technology moves from one entity to another. Success in technology transfer is achieved when the receiving entity, known as the transferee, can efficiently utilize the transferred technology and eventually integrate it into its operations. This transfer may involve the exchange of physical assets, know-how, and technical knowledge. Additionally, technology transfer refers to the movement of technology from laboratory settings to industrial applications, from developed nations to developing ones, or from one application domain to another.

Modes and Mechanisms of Technology Transfer

A. The Bar-Zakay Model

Bar-Zakay (1971) developed a comprehensive model for technology transfer (TT) based on a project management approach. He divided the TT process into four stages: Search, Adaptation, Implementation, and Maintenance. In this model, activities, milestones, and decision points (go or no-go) are depicted for each stage. The upper half of the figure outlines the activities and requirements of the transferor (referred to as the "donor" by Bar-Zakay), while the lower half details those of the transferee or the "recipient." This model specifies the activities to be undertaken in detail and underscores the importance of both the transferor and transferee acquiring skills for technological forecasting, long-range planning, and gathering project-related intelligence.

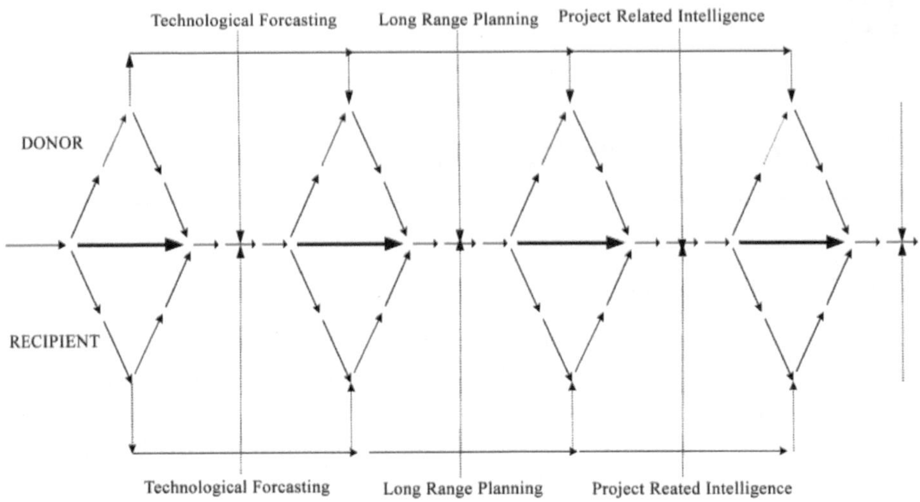

The Bar-Zakay model of technology transfer: Jagoda (2007)

B. The Behrman and Wallender Model

Behrman and Wallender (1976) proposed a seven-stage process for international technology transfer, particularly relevant to multinational corporations. These stages are as follows:

1. *Manufacturing proposal and planning:* Involves decisions regarding location and preparing a business case, including resource assessments.
2. *Deciding product design technologies:* Determining the technologies to be transferred for product design.
3. *Plant design specification:* Detailing the plant design and construction aspects.
4. *Plant construction and production start-up*: Initiating the construction of the plant and starting production.
5. *Process and product adaptation*: Adapting the process and product as needed to suit local conditions.
6. *Product technology improvement*: Enhancing the transferred product technology using local skills.
7. *Providing external support*: Offering external support to strengthen the relationship between the transferor and transferee.

C. The Dahlman and Westphal Model

Dahlman and Westphal (1981), based on their experience in rapidly industrializing countries during the 1980s, proposed a nine-stage process model, which is as follows:

1. *Pre-investment feasibility*: Gathering information and conducting a techno-economic analysis to establish project viability.
2. *Preliminary technology identification*: Identifying technologies needed based on the feasibility study.
3. *Basic engineering studies:* Preparation of process flow diagrams, layouts, and other design specifications.
4. *Detailed engineering study:* Preparing detailed civil engineering plans and identifying peripheral technology.
5. *Supplier selection:* Selecting suppliers for equipment and subcontracting services.
6. *Training and education:* Developing a plan for training workers involved in the technology transfer project.
7. *Plant construction:* Constructing the plant.
8. *Commencing operations:* Starting operations.
9. *Troubleshooting and problem-solving:* Developing skills to address design and operational problems as they arise.

D. The Schlie, Radnor, and Wad Model

Schlie et al. (1987) proposed a generic model comprising seven elements influencing the planning, implementation, and success of any technology transfer project:

1. *Transferor*: The entity selling the technology.
2. *Transferee*: The entity buying the technology.
3. *Technology*: The technology being transferred.
4. *Transfer mechanism:* The method chosen to transfer the technology.
5. *Transferor environment:* Conditions influencing the transferor's operations.
6. *Transferee environment:* Conditions affecting the transferee's absorptive capacity.

7. *Greater environment:* Surrounding conditions, including political relationships and economic factors.

These elements remain relevant in today's business environment, though their manifestations may change over time.

E. The Chantramonklasri Model

Chantramonklasri (1990) further improved the Dahlman and Westphal Model by proposing a five-phase model:

1. Pre-investment and feasibility study.
2. Engineering specifications and design development.
3. Commencing capital goods production based on developed designs.
4. Commissioning and start-up, including comprehensive workforce training.
5. Commencing commercial production.

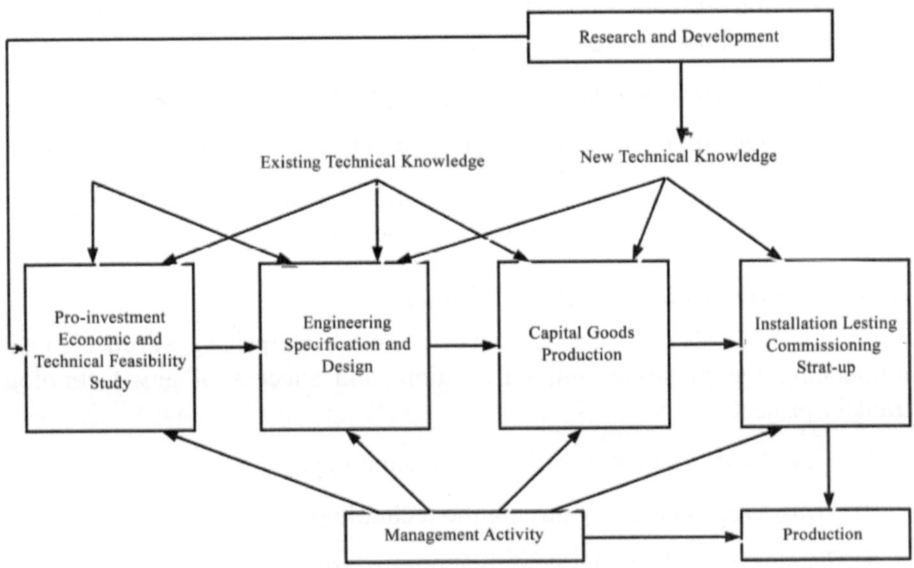

The Five-phase model of international technology transfer Source: Jagoda (2007)

A. The Appropriability Model

The Appropriability Model, originating in the 1945-1950s, posits that high-quality technologies inherently attract users without the need for deliberate technology transfer mechanisms. It underscores the significance of research quality and market competitiveness in facilitating technology transfer. According to this model, technology transfer occurs naturally as users recognize and adopt available technologies. The deliberate promotion of technology transfer mechanisms is deemed unnecessary, as it is assumed that once researchers develop technologies and disseminate them through technical reports and professional journals, users will readily engage with them.

B. The Dissemination Model

Popularized by Rogers (1983) and Rogers and Kincaid (1982) in the 1960-1970s, the Dissemination Model emphasizes the diffusion of technology and innovation from experts to potential users. It posits that experts transfer specialized knowledge to willing users, leading to the spread of new technologies. The underlying assumption is that once established, the linkage between experts and non-experts facilitates the smooth flow of technology, akin to water through a pipe once the channel is opened.

C. The Knowledge Utilization Model

Developed in the late 1980s, the Knowledge Utilization Model focuses on the interpersonal communication between technology developers/researchers and users, as well as the organizational factors that either hinder or facilitate technology transfer. This approach emphasizes organizing knowledge effectively for practical application in the users' context, representing a step forward in understanding how knowledge can be utilized efficiently.

D. The Communication Model

The Communication Model views technology transfer as a process of communication and information flow, characterized by the exchange and sharing of meanings. It suggests that technology transfer involves continuous and simultaneous interaction among individuals, with knowledge being perceived as an objective entity waiting to be communicated. The implementer (technology developer) is responsible for conveying knowledge accurately through appropriate channels, and any failure in adoption is attributed to users' failure to comprehend.

E. Gibson and Slimor's Model

Gibson and Slimor's Model conceptualizes technology transfer through three levels of involvement: Technology Development, Technology Acceptance, and Technology Application. Drawing from organization and communication theories, this model integrates activities from traditional models. Technology Development, akin to the Appropriability Model, highlights the passive transfer of technology through means such as research reports and journal articles, emphasizing the importance of research quality and market competitiveness.

Capacity Building of Extension Personnel

Capacity Building encompasses activities aimed at enhancing the knowledge, abilities, skills, and behavior of individuals, as well as improving institutional structures and processes to efficiently achieve organizational missions and goals sustainably. Training serves as a fundamental component of capacity building.

19

Training for Development

Training

Training refers to the systematic process of imparting knowledge, skills, and competencies to individuals or groups to enhance their performance, capabilities, and effectiveness in a specific area or field. It involves structured learning activities designed to achieve predetermined learning objectives and outcomes. Training can occur through various modalities, including classroom instruction, hands-on practice, on-the-job training, workshops, seminars, and e-learning platforms.

The term “training” has been in use for centuries, originating from the Old English word “trainian,” meaning “to draw or drag” or “to make prepared.” Over time, the term evolved to refer to the process of preparing or instructing individuals for specific tasks or roles.

Several famous authors and scholars have provided definitions of training:

Edgar Dale: Dale defined training as "the organized activity aimed at imparting information and/or instructions to improve the recipient's performance or to help him attain a required level of knowledge or skill."

Raymond Noe: Noe defined training as "a systematic process of providing employees with the competencies (knowledge, skills, and abilities) needed for their current or future jobs."

Robert L. Cannon: Cannon defined training as "a planned process to modify attitude, knowledge, skill, or behavior through learning experience to achieve effective performance in an activity or range of activities."

These definitions highlight key elements of training, including its systematic nature, its focus on improving performance or achieving specific outcomes, and its emphasis on learning and skill development. Training plays a critical role in enhancing individual and organizational capabilities, improving productivity and efficiency, and driving organizational success and innovation.

Training V/S Education

S. No	Training	Education
1.	Training is meant for those in the jobs or vocation	Education is mostly attended by fresh and inexperienced learners
2.	The learners are adult and from diverse background	Learners are similar in background and characteristics
3.	Training is generally organised forshorter duration and for specific purpose in view	Education is long term affair
4.	Training courses are planned on the basis of needs of the participants and thus are quite focused and flexible in nature	Education is broad and cateringto general needs of people. Curricula are rigid and take longer time to update and change
5.	Training aims to improve performance of people. Thus there isa sense of urgency about providing ability to use all that is taught during training	Education is mostly concerned with enhancing knowledge. Even though practical may be part of education, most of what is taught is seldom is put top practice

Teacher V/ S Trainer

S. No.	Teacher	Trainer
1.	Teacher works as experts. He seems to know all the thingsand learners are mostly ignorant	Trainer acts as co-learners. He is one among other knowledgeable and experienced people. Thus, he shares ideas, inspires and help them learn together
2.	Teacher as authority in theclass and control the learners. He maintain the disciplines	Trainer is mostly non-interfering. He tries to create a positive learning climate in order to encourage learners to assume responsibility
3.	Teachers is generally uselectures	Trainers is generally use a variety of teaching methods to create interests and involvement of learners
4.	Teacher expects respect and obedience from learners	Trainer respects ability and experience of the learners
5.	Teachers inform plans ofteaching and expects the learners follow	Trainers presents plan of training and invites comments from learners to make suitable changes

Features of Training

The features of training encompass various characteristics and attributes that define the nature and effectiveness of the learning process. Here are some key features of training:

1. *Structured Learning:* Training involves a systematic and organized approach to learning, with clearly defined objectives, content, and activities designed to achieve specific learning outcomes.
2. *Goal-Oriented:* Training is focused on achieving predetermined goals and objectives, such as acquiring new knowledge, developing skills, changing behaviors, or improving performance in a particular area or field.
3. *Participant-Centered:* Effective training programs are designed with the needs, preferences, and capabilities of participants in mind. They are tailored to the specific audience, taking into account factors such as prior knowledge, learning styles, and individual learning goals.
4. *Interactive and Engaging*: Training activities should be interactive, engaging, and participatory to promote active learning and retention of information. They may include group discussions, case studies, role-playing exercises, simulations, and hands-on practice.
5. *Application-Oriented*: Training emphasizes the practical application of knowledge and skills to real-world situations and tasks relevant to participants' roles or responsibilities. It seeks to bridge the gap between theory and practice by providing opportunities for hands-on learning and skill development.
6. *Continuous and Lifelong*: Training is an ongoing process that extends beyond initial learning experiences. It involves continuous learning, skill enhancement, and professional development throughout one's career or lifetime to adapt to changing job requirements, technologies, and industry trends.
7. *Feedback and Assessment:* Training programs incorporate mechanisms for providing feedback and assessing participants' progress and performance. Feedback helps learners identify areas for improvement and reinforces learning, while assessment measures the attainment of learning objectives and informs future training efforts.
8. *Flexibility and Adaptability*: Effective training programs are flexible and adaptable to accommodate diverse learning needs, preferences, and contexts. They may incorporate various delivery methods, formats, and technologies to cater to different learning styles and preferences.
9. *Evaluation and Improvement*: Training programs include processes for evaluating their effectiveness and impact, such as post-training

assessments, participant feedback surveys, and performance evaluations. Evaluation findings are used to identify strengths, weaknesses, and areas for improvement to enhance the quality and relevance of future training efforts.

10. *Ethical and Inclusive*: Training adheres to ethical principles and values, such as respect, integrity, and fairness, in its design, delivery, and evaluation. It promotes inclusivity and diversity, ensuring equal access and opportunities for all participants regardless of background, gender, ethnicity, or other characteristics.

These features collectively contribute to the effectiveness, relevance, and impact of training programs in enhancing individual and organizational performance, promoting continuous learning and development, and driving success and innovation.

Phases of training

The three important phases of training are:

1. Pre-Training Phase

- *Needs Assessment:* This phase involves identifying the training needs of individuals or groups within an organization. It may include conducting surveys, interviews, or performance evaluations to determine gaps in knowledge, skills, or competencies.
- *Setting Objectives:* In this phase, clear and specific learning objectives are established based on the identified training needs. Objectives should be SMART (Specific, Measurable, Achievable, Relevant, and Time-bound) to guide the design and evaluation of the training program.
- *Planning and Preparation:* During this phase, training content, methods, materials, and resources are developed and organized according to the established objectives. A training plan is created, outlining the sequence of activities, timeline, and logistics for implementation.

2. Training Delivery Phase

- Implementation: This phase involves delivering the training program according to the planned schedule and format. Trainers conduct sessions using various instructional methods, such as lectures, discussions, demonstrations, hands-on activities, and simulations, to engage participants and facilitate learning.

- *Facilitation and Support:* Trainers provide guidance, support, and assistance to participants throughout the training process. They answer questions, clarify concepts, and address concerns to ensure that learners understand the material and can apply it effectively.
- *Assessment and Feedback:* Assessment activities, such as quizzes, tests, assignments, or practical exercises, are conducted to evaluate participants' understanding and mastery of the training content. Feedback is provided to participants to reinforce learning, identify areas for improvement, and guide further practice or study.

3. Post-Training Phase

- *Evaluation*: This phase involves assessing the effectiveness and impact of the training program. Evaluation may include gathering feedback from participants, trainers, and stakeholders, as well as measuring changes in knowledge, skills, attitudes, or behaviors resulting from the training.
- *Follow-Up and Reinforcement*: Follow-up activities are conducted to support participants in applying their newly acquired knowledge and skills in their work or daily activities. This may include providing additional resources, mentoring, coaching, or on-the-job support to reinforce learning and facilitate transfer of training.
- *Continuous Improvement*: Evaluation findings are used to identify strengths, weaknesses, and areas for improvement in the training program. Lessons learned are incorporated into future training efforts to enhance effectiveness, relevance, and impact.

These three phases are essential components of the training process and contribute to its success in achieving desired learning outcomes and performance improvements.

Three Basic Phase in Learning

Kurt Lewin, a pioneering psychologist known for his work in social psychology and organizational behavior, proposed a three-phase model of change known as the "unfreezing-change-refreezing" model. While this model is primarily focused on organizational change, it can also be applied to learning situations. The three basic phases in learning, as adapted from Lewin's model, are:

1. Unfreezing

- In the unfreezing phase, learners become aware of the need for change and are motivated to overcome resistance to learning. This phase involves creating a sense of psychological safety and readiness for learning by challenging existing beliefs, attitudes, and behaviors.
- Learners may experience discomfort or dissonance as they confront their current assumptions and recognize the need to acquire new knowledge, skills, or perspectives. Unfreezing involves breaking down existing mental barriers and opening the mind to new possibilities.

2. Change

- In the change phase, learners engage in the process of acquiring new knowledge, skills, or behaviors. This phase involves active participation in learning activities, such as attending training sessions, practicing new skills, seeking feedback, and reflecting on learning experiences.
- Learners experiment with new ideas, strategies, and approaches, gradually integrating them into their existing knowledge and practice. Change involves cognitive and behavioral shifts as learners adopt new ways of thinking and acting in response to the learning context.

3. Refreezing

- In the refreezing phase, learners solidify and integrate their new learning into their everyday practice. This phase involves reinforcing and institutionalizing the changes to ensure their long-term sustainability and effectiveness.
- Learners seek opportunities to apply their new knowledge and skills in real-world situations, receive reinforcement and recognition for their efforts, and establish new habits and routines. Refreezing involves consolidating learning gains and embedding them into the individual's or organization's culture, processes, and systems.

These three phases—unfreezing, change, and refreezing—constitute a cyclical process of learning and change. While the model was originally developed to explain organizational change, it can also be applied to individual and group learning contexts, highlighting the importance of motivation, engagement, and reinforcement in the learning process.

Types of Training

There are various types of training, each designed to meet specific learning objectives, accommodate different learning styles, and address diverse training needs. Here's an explanation of some common types of training:

On-the-Job Training (OJT)

- *Description*: On-the-job training involves learning while performing tasks or duties in the actual work environment. It is typically conducted by experienced colleagues or supervisors who provide guidance, instruction, and hands-on demonstration of job tasks.
- *Application*: OJT is commonly used for teaching job-specific skills, procedures, and techniques to new employees or those transitioning to new roles within an organization. It allows learners to gain practical experience and apply theoretical knowledge in real-world settings.

Classroom Training

- *Description*: Classroom training, also known as instructor-led training (ILT), involves face-to-face instruction conducted in a classroom or training facility. It is facilitated by an instructor or trainer who delivers lectures, presentations, demonstrations, and interactive activities to a group of learners.
- *Application*: Classroom training is suitable for teaching theoretical concepts, foundational knowledge, and complex topics that require explanation, discussion, and interaction among participants. It allows for structured learning, peer collaboration, and immediate feedback from the instructor.

E-Learning

- *Description*: E-learning, or online training, involves the use of digital technologies and internet-based platforms to deliver educational content and facilitate learning remotely. It encompasses various formats, including webinars, video lectures, interactive modules, quizzes, and discussion forums.
- *Application*: E-learning is widely used for providing flexible and accessible training to geographically dispersed learners, remote workers, or individuals with busy schedules. It allows learners to progress at their own pace, access learning materials anytime, anywhere, and engage with multimedia content.

Hands-On Training

- *Description*: Hands-on training involves experiential learning through practical exercises, simulations, and role-playing activities. Learners actively engage in performing tasks, manipulating equipment, and solving problems to gain proficiency and confidence in their skills.
- *Application*: Hands-on training is effective for teaching technical skills, operational procedures, and troubleshooting techniques that require physical practice and application. It provides a safe and controlled environment for learners to develop competence and mastery through trial and error.

Mentoring and Coaching

- *Description*: Mentoring and coaching involve one-on-one guidance and support provided by experienced professionals (mentors or coaches) to less experienced individuals (mentees or coachees). Mentors and coaches offer personalized feedback, advice, and encouragement to help mentees achieve their learning and development goals.
- *Application*: Mentoring and coaching are beneficial for promoting individual growth, career development, and leadership skills. They provide opportunities for mentees to learn from the expertise and experiences of mentors, receive constructive feedback, and gain insights into their strengths and areas for improvement.

Blended Learning

- *Description*: Blended learning combines multiple training modalities, such as classroom instruction, e-learning modules, and hands-on activities, to create a comprehensive and flexible learning experience. It integrates the strengths of different approaches to accommodate diverse learning preferences and optimize learning outcomes.
- *Application*: Blended learning is suitable for delivering complex or multifaceted training programs that require a combination of instructional methods and resources. It allows organizations to leverage technology, maximize training efficiency, and tailor learning experiences to the needs of individual learners or specific training objectives.

These are just a few examples of the different types of training available. However, the choice of training type depends on factors such as learning objectives, audience characteristics, available resources, and organizational goals. Effective training programs often incorporate a combination of training methods to create engaging, interactive, and impactful learning experiences for participants.

Training provided to farmers and agricultural extension officials encompasses a wide range of topics and methodologies tailored to address specific needs, challenges, and objectives within the agricultural sector. Here are some common types of training given to farmers and agricultural extension officials:

Training for Farmers

1. Technical Skills Training

- *Crop Production Techniques*: Training on best practices for soil preparation, planting, irrigation, fertilization, pest and disease management, and harvesting techniques.
- *Livestock Management*: Training on animal husbandry practices, including feeding, breeding, health care, housing, and disease prevention.
- *Agroforestry and Horticulture*: Training on the cultivation and management of trees, fruits, vegetables, and ornamental plants for food, income, and environmental benefits.
- *Post-Harvest Handling*: Training on proper handling, storage, processing, and marketing of agricultural products to minimize losses and maximize value.

2. Technology Adoption and Innovation

- *Introduction to New Technologies*: Training on the use of modern agricultural technologies, such as improved seeds, mechanization, precision agriculture, greenhouse technologies, and biotechnology.
- *ICT and Mobile Applications*: Training on the use of information and communication technologies (ICT), mobile apps, and digital platforms for accessing market information, weather forecasts, extension services, and financial services.

3. Sustainable Agriculture and Natural Resource Management

- *Organic Farming*: Training on organic farming practices, including composting, crop rotation, mulching, and biological pest control, to reduce reliance on synthetic inputs and promote environmental sustainability.

- *Water Management*: Training on water conservation techniques, efficient irrigation methods, rainwater harvesting, and watershed management to optimize water use and protect water resources.
- *Soil Conservation*: Training on soil conservation practices, such as contour plowing, terracing, cover cropping, and agroforestry, to prevent erosion, improve soil health, and enhance land productivity.

4. Business and Financial Management

- *Farm Business Planning*: Training on farm business planning, budgeting, recordkeeping, and financial analysis to improve farm profitability, decision-making, and risk management.
- *Marketing and Value-Chain Development*: Training on market analysis, value addition, packaging, branding, and access to markets and market information to increase farm income and competitiveness.
- *Microfinance and Entrepreneurship*: Training on accessing credit, savings, insurance, and other financial services, as well as entrepreneurship skills development, to support farm diversification and rural enterprise development.

Training for Agricultural Extension Officials

1. Extension Methodologies and Approaches

- *Adult Learning Principles*: Training on principles of adult learning and participatory approaches to extension, including farmer field schools, demonstration plots, group discussions, and experiential learning methods.
- *Communication and Facilitation Skills*: Training on effective communication, facilitation, and interpersonal skills for engaging with farmers, building trust, and fostering collaborative relationships.

2. Technical and Subject Matter Training

- *Technical Crop and Livestock Production*: Training on agronomy, horticulture, animal science, and other technical aspects of crop and livestock production to provide accurate and timely advice to farmers.
- *Climate-Smart Agriculture*: Training on climate change adaptation and mitigation strategies, including climate-smart agricultural practices, resilient crop varieties, and weather forecasting tools.

3. ICT and Digital Extension Tools

- *ICT Literacy*: Training on using information and communication technologies (ICT), mobile applications, and digital platforms for delivering extension services, accessing agricultural information, and communicating with farmers.
- *Extension Software and Databases*: Training on using extension software, databases, and decision-support tools for data management, analysis, reporting, and monitoring and evaluation (M&E).

4. Gender and Social Inclusion

- *Gender-Sensitive Extension*: Training on gender-sensitive approaches to extension, including understanding gender dynamics, addressing gender-based constraints, and promoting women's participation and empowerment in agriculture.
- *Inclusive Extension*: Training on inclusive approaches to extension, including reaching out to marginalized groups, such as youth, women, indigenous peoples, and persons with disabilities, and ensuring their participation and access to extension services and resources.

These are just a few examples of the types of training provided to farmers and agricultural extension officials. Training programs are often customized to meet the specific needs and priorities of target audiences, taking into account factors such as location, agroecological conditions, socio-economic context, and cultural preferences. Effective training plays a critical role in building the capacity of farmers and extension professionals, promoting sustainable agriculture, and contributing to food security and rural development.

Methods of training

Methods of training refer to the various approaches and techniques used to deliver educational content and facilitate learning. These methods are chosen based on the learning objectives, target audience, available resources, and training context. Here's an overview of some common methods of training:

Lectures

Description: Lectures involve a one-way communication approach where an instructor presents information to learners through verbal communication, often accompanied by visual aids such as slides or presentations.

Application: Lectures are effective for delivering large amounts of information to a group of learners in a relatively short period. They are suitable for introducing new concepts, theories, and principles, as well as providing an overview of complex topics.

Demonstrations

Description: Demonstrations involve showing learners how to perform a task or operate equipment through practical examples or step-by-step instructions.

Application: Demonstrations are effective for teaching hands-on skills and procedures, such as equipment operation, laboratory techniques, or assembly processes. They provide visual reinforcement and clarification of concepts and techniques.

Hands-On Practice

Description: Hands-on practice allows learners to actively engage in performing tasks, applying knowledge, and practicing skills in a controlled environment.

Application: Hands-on practice is essential for skill-based training, such as technical skills, manual tasks, or procedures. It provides opportunities for learners to gain practical experience, build confidence, and reinforce learning through repetition and feedback.

Case Studies

Description: Case studies present real-life scenarios or problems for analysis and discussion, allowing learners to apply theoretical knowledge to practical situations.

Application: Case studies are effective for promoting critical thinking, problem-solving, and decision-making skills. They encourage active participation, collaboration, and reflection among learners and provide insights into real-world challenges and solutions.

Simulations

Description: Simulations replicate real-world environments, processes, or situations to provide learners with immersive learning experiences.

Application: Simulations are useful for training in high-risk or complex environments, such as emergency response, aviation, healthcare, or military operations. They allow learners to practice skills, make decisions, and experience consequences in a safe and controlled setting.

Role-Playing

Description: Role-playing involves learners assuming different roles or characters to enact scenarios or interactions, simulating real-life situations and behaviors.

Application: Role-playing is effective for developing interpersonal skills, communication skills, and empathy. It allows learners to practice communication, negotiation, and conflict resolution in various contexts, such as customer service, sales, or leadership.

Group Discussions

Description: Group discussions involve interactive exchanges of ideas, opinions, and experiences among learners facilitated by an instructor or moderator.

Application: Group discussions promote active participation, collaboration, and knowledge sharing. They encourage critical thinking, problem-solving, and peer learning by allowing learners to explore different perspectives, ask questions, and engage in dialogue.

E-Learning escription: E-learning, or online training, utilizes digital technologies and internet-based platforms to deliver educational content and facilitate learning remotely.

Application: E-learning provides flexible and accessible training options for learners to access learning materials anytime, anywhere. It encompasses various formats, such as webinars, video lectures, interactive modules, quizzes, and discussion forums, catering to diverse learning preferences and needs. These methods of training can be used individually or in combination to create engaging, effective, and comprehensive learning experiences for participants. The choice of training methods depends on factors such as learning objectives, audience characteristics, training goals, and available resources. Effective training programs often incorporate a mix of methods to accommodate different learning styles, promote active engagement, and optimize learning outcomes.

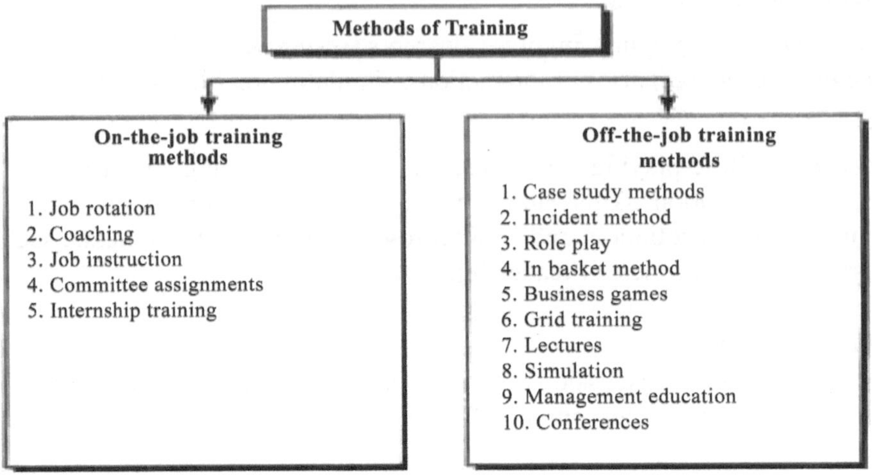

On-the-job training (OJT) method and off-the-job training method are two distinct approaches to delivering training to employees, each with its own characteristics and advantages. Here's an overview of both:

1. On-the-Job Training (OJT) Method

On-the-job training (OJT) involves learning while performing tasks or duties in the actual work environment. It typically takes place at the workplace and is facilitated by experienced colleagues, supervisors, or trainers who provide guidance, instruction, and hands-on demonstration of job tasks.

Characteristics

OJT allows trainees to acquire job-related skills and knowledge through direct experience and practice. Trainees learn in the context of their actual job roles, tasks, and responsibilities, making the training experience highly relevant and applicable to their work. Trainees receive immediate feedback from supervisors or mentors as they perform tasks, enabling them to correct mistakes, reinforce learning, and improve performance in real time. Cost-effective: OJT does not require dedicated training facilities or resources, making it a cost-effective training method for organizations.

Examples: Apprenticeships, job shadowing, mentoring, coaching, rotation assignments, and learning by doing are common examples of on-the-job training methods.

2. *Off-the-Job Training Method:*

Off-the-job training (OJT) involves learning that takes place away from the workplace, often in a formal training setting such as a classroom, training facility, or external training provider. It may involve structured learning activities, lectures, demonstrations, simulations, or hands-on exercises conducted by instructors or trainers.

Characteristics

Off-the-job training provides a structured and controlled learning environment conducive to focused learning and skill development. Trainees receive theoretical knowledge, concepts, and principles relevant to their job roles, which they can apply in practical situations. Off-the-job training methods offer a variety of learning experiences, including lectures, presentations, demonstrations, group discussions, case studies, and simulations. Off-the-job training may require dedicated training facilities, equipment, materials, and resources, which may incur additional costs for organizations.

Examples: Classroom training, workshops, seminars, conferences, e-learning, simulations, role-playing, and case studies are common examples of off-the-job training methods.

Both on-the-job and off-the-job training methods have their advantages and limitations, and the choice between them depends on factors such as the nature of the job, training objectives, learning preferences, available resources, and organizational priorities. Many organizations use a combination of both methods to provide comprehensive and effective training to their employees, leveraging the strengths of each approach to optimize learning outcomes and performance improvement.

Training Need Assessment

Training Needs Assessment (TNA) is a systematic process of identifying the knowledge, skills, competencies, and behaviors required for individuals or groups to perform their jobs effectively. It involves analyzing gaps between the current state of knowledge and skills and the desired state needed to achieve organizational goals and objectives. Here's an overview of the key components and steps involved in training needs assessment:

1. *Identifying Objectives:* The first step in TNA is to clearly define the objectives and goals of the training program. This involves understanding the organizational goals, departmental priorities, and job-specific requirements that training should address.

2. *Data Collection*: TNA involves collecting relevant data and information to assess the training needs of individuals or groups. This may include:
   - Job analysis: Reviewing job descriptions, tasks, and responsibilities to identify the knowledge, skills, and competencies required for job performance.
   - Performance analysis: Analyzing performance metrics, feedback, and evaluations to identify areas of strength and areas needing improvement.
   - *Employee surveys:* Gathering input from employees through surveys, questionnaires, or interviews to understand their training preferences, perceived skill gaps, and learning needs.
   - *Observation*: Observing employees in their work environment to identify specific challenges, obstacles, or areas for improvement.
3. *Analysis and Prioritization:* Once data is collected, it is analyzed to identify training needs and prioritize them based on factors such as:
   - Criticality: Assessing the importance of specific skills or competencies to job performance and organizational success.
   - Frequency: Determining the frequency and prevalence of skill gaps or performance issues across the organization or department.
   - Impact: Estimating the potential impact of addressing specific training needs on individual and organizational performance.
4. *Developing Training Objectives:* Based on the analysis, clear and specific training objectives are developed to address identified training needs. These objectives should be SMART (Specific, Measurable, Achievable, Relevant, and Time-bound) to guide the design and implementation of the training program.
5. *Designing Training Interventions:* Once training objectives are established, appropriate training interventions and methods are selected to address identified needs. This may include on-the-job training, classroom training, e-learning, coaching, mentoring, or other development activities.
6. *Implementation and Evaluation:* The training program is implemented according to the designed plan, and its effectiveness is evaluated through various methods, such as pre- and post-training assessments, participant feedback, performance evaluations, and business impact analysis.

7. *Continuous Improvement:* TNA is an ongoing process, and organizations should continuously monitor, evaluate, and update training programs to ensure they remain aligned with changing organizational goals, job requirements, and employee needs.

By conducting a thorough TNA, organizations can ensure that training efforts are targeted, relevant, and aligned with organizational objectives, resulting in improved employee performance, productivity, and job satisfaction.

Advantages of TNA

Training Needs Assessment (TNA) offers several advantages to organizations, employees, and other stakeholders involved in training and development initiatives. Some of the key advantages include:

1. *Alignment with Organizational Goals:* TNA helps ensure that training efforts are closely aligned with organizational goals, priorities, and strategic objectives. By identifying and addressing skill gaps and performance issues, training programs can contribute directly to enhancing organizational effectiveness and competitiveness.
2. *Targeted Training Investments:* TNA enables organizations to allocate training resources effectively by focusing investments on areas of greatest need. By prioritizing training needs based on their impact and importance, organizations can optimize the use of financial and human resources for training and development activities.
3. *Improved Employee Performance:* By identifying and addressing skill gaps and training needs, TNA helps improve employee performance and productivity. Training programs designed to address specific skill deficiencies or knowledge gaps enable employees to perform their jobs more effectively, leading to enhanced job satisfaction and motivation.
4. *Enhanced Employee Engagement and Satisfaction:* TNA demonstrates an organization's commitment to employee development and career growth, which can boost employee morale, engagement, and job satisfaction. Employees appreciate opportunities for learning and development that are tailored to their needs and contribute to their professional growth.
5. *Reduction in Turnover and Absenteeism:* Effective training programs that address employee skill gaps and career development needs can help reduce turnover rates and absenteeism. Employees who receive relevant training and support are more likely to feel valued and engaged, leading to higher levels of job satisfaction and retention.

6. *Identification of Training Priorities:* TNA helps organizations identify and prioritize training needs based on their impact on job performance and organizational success. By focusing on critical skill areas and competency gaps, organizations can address immediate training needs while also planning for long-term skill development and succession planning.
7. *Optimized Training Program Design:* TNA provides valuable insights into the learning preferences, needs, and challenges of employees, enabling organizations to design training programs that are tailored to their specific requirements. By incorporating input from employees and stakeholders, organizations can create engaging, relevant, and effective training experiences.
8. *Enhanced Organizational Agility and Adaptability:* By continuously assessing training needs and investing in employee development, organizations can build a more agile and adaptable workforce capable of responding to changing market conditions, technological advancements, and industry trends. Training programs that focus on building transferable skills and competencies enable employees to adapt to new roles and responsibilities more effectively.

Overall, Training Needs Assessment (TNA) plays a critical role in driving organizational success by ensuring that training efforts are targeted, relevant, and aligned with organizational objectives and employee needs.

Approaches of TNA

There are several approaches for conducting Training Needs Assessment (TNA), each with its own methodologies, techniques, and applications. The choice of approach depends on factors such as the organization's goals, the nature of the training needs, available resources, and the preferences of stakeholders. Here are some common approaches for TNA:

1. Organizational Analysis

- Organizational analysis focuses on identifying training needs at the organizational level by examining strategic goals, performance indicators, and operational challenges. It involves analyzing organizational structure, culture, processes, and systems to understand how they impact training requirements.
- Review of organizational goals and objectives, analysis of performance metrics and key performance indicators (KPIs), interviews with senior

management, surveys of employees, examination of organizational policies and procedures.

- Organizational analysis helps identify overarching training needs related to organizational strategy, leadership development, change management, and workforce planning.

2. Task Analysis

- Task analysis involves breaking down job roles and tasks into specific activities, skills, and competencies required for successful job performance. It focuses on understanding the knowledge, skills, and abilities (KSAs) needed to perform job tasks effectively.
- Job shadowing, observation, interviews with subject matter experts (SMEs), job analysis questionnaires, work sampling, review of job descriptions and performance standards.
- Task analysis helps identify job-specific training needs related to technical skills, procedural knowledge, safety procedures, and job-related competencies.

3. Individual Analysis

- Individual analysis examines the performance of individual employees to identify skill gaps and training needs. It focuses on assessing the knowledge, skills, abilities, and performance levels of individual employees relative to job requirements.
- Performance appraisals, skills assessments, competency assessments, self-assessments, 360-degree feedback, interviews with supervisors and employees.
- Individual analysis helps identify training needs related to specific employees' performance issues, career development goals, learning preferences, and skill deficiencies.

4. Work Environment Analysis

- Work environment analysis examines the organizational culture, work climate, physical environment, and support systems that impact employee performance and learning. It focuses on identifying barriers and facilitators to learning within the work environment.
- Surveys of employees, focus groups, workplace observations, analysis of work conditions and facilities, review of support systems and resources.

- Work environment analysis helps identify environmental factors that may affect employee motivation, engagement, and performance, such as lack of resources, communication barriers, or cultural norms.

5. Stakeholder Analysis

- Stakeholder analysis involves identifying and engaging stakeholders who have a vested interest in training and development initiatives. It focuses on understanding their perspectives, needs, expectations, and concerns regarding training.
- Stakeholder interviews, focus groups, surveys, stakeholder mapping, social network analysis.
- Stakeholder analysis helps ensure that training initiatives are aligned with stakeholder interests and priorities, fostering stakeholder buy-in, support, and collaboration throughout the TNA process.

By using a combination of these approaches, organizations can conduct a comprehensive Training Needs Assessment (TNA) to identify and prioritize training needs effectively. Each approach provides valuable insights into different aspects of training requirements, enabling organizations to design and implement targeted, relevant, and impactful training programs.

20

Human Resource Development

The concept of Human Resource Development (HRD) originated at George Washington University in the USA and gained prominence during the American Society for Training and Development conference in Miami in 1969. HRD aims to enhance individual awareness of their organizational significance and provide ample opportunities for effective functioning, thereby empowering individuals. It involves the establishment of organizational systems and activities to foster employee motivation, responsibility, and holistic development.

In the realm of extension services, effective planning and management of human resources are imperative to enhance the capabilities, motivation, and overall efficacy of extension personnel. To foster a self-reliant approach to sustainable agriculture, extension personnel require training to sensitively address the diverse needs of farmers.

Definition

Rao (1992) *explained that HRD is a process in which employees of an organization are continuously helped in planned manner to acquire and sharpen capabilities that are required to performed various function.*

Nadler (1994) *defined HRD as a planned and continuous effort by management to improve employee's capacity levels and organizational performance through training, education and develop programmes. Training includes those activities that are conducted to improve overall competence of an individual in a specific present or future job.*

The Extension Committee on Policy in the USA identified eight key areas for training Extension agents:

1. Extension organization and administration
2. Programme planning and development
3. Communication
4. Research

5. Human development
6. Educational processes
7. Social processes
8. Effective thinking

Core extension personnel in developing countries typically include Village Extension Workers (VEW), Subject Matter Specialists (SMS), and Supervisory Staff or Extension Officers.

Rolling (1988) elucidated the HRD approach at the grassroots level through five functions:

1. Mobilization of people
2. Organization
3. Technical support
4. Training
5. System management

Emerging areas of training include

1. Training in new agricultural technologies
2. Training in sustainable agriculture practices
3. Training to enhance farmers' capacity for self-development
4. Training programs tailored for youth and female farmers

Infrastructure for extension training is established at the national level, spearheaded by the Directorate of Extension under the Ministry of Agriculture and Cooperation, Government of India. This involves collaborative planning of training courses with state agricultural universities, national institutes, ICAR institutes, State departments of agriculture, and other relevant institutions. A network of training institutions operates at various levels, including national, regional, state, divisional, district, and block levels. Key institutions include:

1. *MANAGE, Hyderabad:* Focuses on training senior extension personnel in research-extension linkage, monitoring, and evaluation.
2. *State Agricultural Universities (SAUs):* Offers training courses for middle-level extension personnel.
3. *Agricultural Colleges and Training Centers (ACTs):* Provides training for middle and senior-level extension personnel.

4. *Extension Education Institutes (EEIs):* Imparts training on extension education concepts, communication media, and training methodology for middle-level extension personnel.
5. *State Institutes of Rural Development (SIRD):* Engaged in rural development training.
6. *Farm Training Centers (FTCs):* Established in the 1960s to train farmers in new agricultural technologies.
7. *Krishi Vigyan Kendra (KVKs):* Focuses on providing need-based skill-oriented training to farmers, rural youths, farm women and grassroot extension functionaries.

21

Extension Teaching Methods

Extension teaching methods play a crucial role in disseminating knowledge and fostering behavioral change among target audiences in agricultural extension programs. These methods encompass a variety of strategies designed to engage, educate, and empower farmers, rural communities, and extension personnel.

> "Extension teaching method s are devices used to create situations in which communication can take place between an instructor and the learners. The basic objective of extension teaching method is to create an opportunity for effective learning in order to secure changes in the minds and action of the learners."
>
> Leagans (1960)

This chapter explores various extension teaching methods, their applications, advantages, and limitations.

1. Lecture Method

The lecture method is a traditional yet effective approach wherein an expert delivers information to a group of learners. It is commonly used to introduce new concepts, present technical information, and provide updates on agricultural practices. Lectures can be supplemented with visual aids such as slides, charts, and multimedia presentations to enhance comprehension and retention. While this method allows for the dissemination of large amounts of information efficiently, it may lack interactivity and fail to address individual learning needs.

2. Demonstrations

Demonstrations involve practical, hands-on sessions where extension agents showcase agricultural techniques, tools, and practices directly to farmers. This method enables participants to observe, learn, and replicate demonstrated skills in their own contexts. Demonstrations are particularly effective for teaching practical skills such as crop planting, pest management, and livestock care. They promote experiential learning and foster confidence among farmers, leading to increased adoption of recommended practices.

3. Field Visits

Field visits provide opportunities for farmers to observe real-life examples of successful agricultural practices in neighboring farms or research stations. Extension agents facilitate discussions, answer questions, and provide guidance during these visits. Field visits enhance farmers' understanding of local conditions, challenges, and opportunities, thereby promoting context-specific learning and problem-solving skills. However, logistical constraints such as transportation and scheduling may limit the frequency and accessibility of field visits.

4. Group Discussions

Group discussions encourage active participation, collaboration, and knowledge sharing among participants. Extension agents facilitate discussions on specific topics or challenges, allowing farmers to share their experiences, insights, and solutions. Group discussions foster peer learning, networking, and community cohesion, empowering farmers to collectively address common issues. However, effective moderation is essential to ensure inclusive participation and constructive dialogue.

5. Participatory Approaches

Participatory approaches such as participatory rural appraisal (PRA), participatory learning and action (PLA), and farmer field schools (FFS) prioritize active involvement and decision-making by farmers throughout the learning process. These methods emphasize collaborative problem identification, experimentation, and evaluation of alternative practices. Participatory approaches promote farmer empowerment, ownership, and sustainability of extension interventions. However, they require significant time, resources, and facilitation skills to implement effectively.

6. Information and Communication Technologies (ICTs)

ICTs, including mobile phones, internet platforms, and multimedia resources, are increasingly used to deliver extension services and educational materials to remote and underserved communities. ICTs enable timely access to relevant information, interactive learning modules, and expert advice. They overcome geographical barriers and facilitate continuous learning beyond traditional extension settings. However, digital literacy, connectivity issues, and infrastructural constraints may limit the reach and effectiveness of ICT-based extension methods.

Extension teaching methods encompass a diverse range of approaches tailored to the needs, preferences, and contexts of target audiences. By combining traditional and innovative methods, extension agents can effectively engage farmers, promote knowledge uptake, and facilitate sustainable agricultural development. Continuous evaluation and adaptation of teaching methods are essential to enhance their relevance, accessibility, and impact in agricultural extension programs.

Classification of Extension Teaching Methods by Wilson & Gallup (1955)

According to Use

Individual Contact Method	Group Contact Method	Mass Contact Method
- Farm & Homevisit - Personal letters - Office call - Flag method - Agriculturalclinic	- Group meeting - Method demonstration - Result demonstration - Conducted tour - Training camps, - Cropschool - Peripatetic team meeting - Media forum - Agricultural games	**a. Broadcast media** - Radio and recordings - Educational television **b. Printed media** - Farm journal - Extension pamphlets - Bulletin, Leaflets - Circular letters **c. Screen media** - Slide, Film, Stripes, - Movies, Videorecordings **d. Others** - Exhibition - Campaign - Farmers fairs

According to Form

Written	Spoken	Visual	A-V or Spoken and Visual
- Bulletin - Leaflets - Personalletters - Circularletters - Farm journal	- Meeting - Farm andHome visit - Office call - Radio & Recordings - Flag method - Agriculturalclinic	- Result demonstration - Exhibition - Poster - Chart - Slides, Filmstripes - Flash cards - Flannel graph - Bulletin boards	- Method demonstration - Result demonstration - Television - Movies - Puppets - Campaign

Classification of Audio-Visual Aids

Audio-visual aids are valuable tools used to enhance the learning experience by combining auditory and visual stimuli. They serve various purposes, from illustrating complex concepts to facilitating interactive discussions.

Audio Aids	**Visual Aids**		**Audio-Visual Aids**
- Radio talk - Tape recording - Disk recording	**Two-dimensional visual**	**Three-dimensional visual**	- Motionpicture - Telecasts - Video recordings - Sound synchronizedslides - Radio-vision
	a. Projected - Films - Slides - Film strips - Opaque material - Overhead transparencies **b. Non-projected Presentation** - Flash card - Flannel graph - Charts - Flip book - Maps **c. Display** - Bulletin boards - Photograph - Posters - Wall chart - Exhibits **d. Literature** - Illustrated - Leaflets - Comics	- Models - Specimens and Sample - Mock-ups - Puppets - 3-D films	

Here, we classify audio-visual aids based on their characteristics and modes of presentation:

1. Visual Aids

Visual aids primarily rely on visual stimuli to convey information. They include

- *Charts and Graphs:* Visual representations of data or information, such as bar graphs, pie charts, and line graphs, to aid comprehension and analysis.
- *Diagrams and Illustrations:* Visual diagrams and drawings depicting processes, structures, or concepts to enhance understanding and retention.
- *Maps*: Geographic representations used to illustrate locations, spatial relationships, and distribution patterns.

- *Photographs and Images:* Visual depictions of real-life scenes, objects, or phenomena to provide context and visual appeal.
- *Models and Objects:* Three-dimensional representations or physical objects used to demonstrate concepts, prototypes, or tangible examples.

2. Audio Aids

Audio aids rely on auditory stimuli, such as sound or speech, to convey information. They include:

- *Recordings:* Pre-recorded audio materials, including lectures, interviews, discussions, or sound effects, played through audio devices such as speakers or headphones.
- *Podcasts*: Digital audio files, typically in episodic format, covering a wide range of topics and accessible through online platforms or mobile applications.
- *Sound Effects*: Auditory cues or effects used to create atmosphere, illustrate scenarios, or enhance engagement during presentations or demonstrations.

3. Audio-Visual Aids

Audio-visual aids combine both auditory and visual elements to create a multimedia learning experience. They include:

- *Slideshows*: Sequential presentations of images, text, and audio narration using slide projection software or multimedia platforms.
- *Videos*: Recorded visual sequences accompanied by sound or narration, ranging from short clips to full-length documentaries, accessible through various platforms such as DVDs, streaming services, or online repositories.
- *Animations*: Dynamic visual representations created through computer graphics or hand-drawn techniques, often used to illustrate processes, simulations, or conceptual models.
- *Interactive Multimedia*: Digital applications or presentations that allow users to engage with content through interactive features such as quizzes, simulations, or virtual tours.

4. Hybrid Aids

Hybrid aids incorporate elements of both traditional and digital formats, offering versatility and adaptability. They include:

- Interactive Whiteboards: Digital display boards that combine touchscreen functionality with multimedia capabilities, enabling real-time annotation, manipulation, and integration of various media formats.
- Augmented Reality (AR) and Virtual Reality (VR): Immersive technologies that overlay digital content onto real-world environments (AR) or create simulated environments (VR) to enhance learning experiences through interactive exploration and engagement.

The classification of audio-visual aids provides a framework for understanding the diverse range of tools available for educational purposes. By selecting and utilizing appropriate aids based on the learning objectives, audience preferences, and technological capabilities, educators can create dynamic and engaging learning environments that maximize comprehension, retention, and overall learning outcomes.

22

ICT Application in Transfer of Technology

The term Information and Communication Technologies (ICT) encompasses a broad spectrum of electronic tools that converge to facilitate information processing, communication, transmission, and display. Scholars like Michiels and Van Crowder (2001) define ICTs as a constellation of electronic technologies capable of transforming organizations and reshaping social relations. In the context of agriculture, the integration of ICT holds significant promise, offering a myriad of advantages ranging from accelerating agricultural growth to empowering marginalized farming communities. This review examines the need, catalytic role, advantages, scope, and limitations of ICT in agricultural extension, elucidating its potential and challenges in promoting sustainable rural development.

Need for ICT in Agriculture Extension

1. *Accelerating Agricultural Growth*: ICTs play a pivotal role in catalyzing agricultural development by providing timely access to critical information, knowledge, and resources.

2. *Expanding Knowledge Resources*: ICTs serve as repositories of agricultural knowledge, enabling extension workers and farmers to access comprehensive information on best practices, innovations, and market trends.

3. *Facilitating Information Services*: ICTs enhance the dissemination of agricultural information and advisory services, enabling farmers to make informed decisions on crop management, pest control, and market opportunities.

4. *Supplementing Technical Manpower*: In resource-constrained settings, ICTs complement the scarcity of technical expertise by providing virtual consultation, training, and technical support to farmers and extension workers.

5. *Strengthening Research-Extension Linkages*: ICTs bridge the gap between agricultural research institutions and extension agencies, facilitating the seamless transfer of research findings, technologies, and innovations to farmers.
6. *Developing Feedback Mechanisms*: ICT platforms enable real-time feedback and interaction between farmers, extension agents, and policymakers, fostering participatory decision-making and adaptive management.
7. *Cost-Effective Extension Delivery*: ICTs offer cost-effective solutions for delivering extension services, reducing the logistical and operational costs associated with traditional outreach methods.
8. *Developing Knowledge Managers*: ICTs empower farmers and extension workers to become knowledge managers, enabling them to curate, share, and apply agricultural information effectively.
9. *Ensuring Gender Equity:* ICT interventions promote gender equity in technology transfer by providing equal access to information, resources, and decision-making platforms for women farmers.
10. *Empowering Smallholders*: ICTs empower small and marginal farmers by providing access to market information, financial services, and agricultural advisory support, thereby enhancing their productivity and resilience.

Catalytic Role and Importance of ICT

1. *Decision-Making Process*: ICTs facilitate evidence-based decision-making by providing timely and relevant information on crop selection, input usage, and market dynamics.
2. *Advisory Services:* ICT platforms offer personalized advisory services tailored to farmers' specific needs, including pest management, soil health, and climate-resilient practices.
3. *Market Outlook:* ICTs provide real-time market information, price trends, and value chain analysis, enabling farmers to optimize their production and marketing strategies.
4. *Empowering Rural Community*: ICT interventions empower rural communities by enhancing their access to education, healthcare, financial services, and market opportunities.

5. *Multisectoral Services:* ICTs promote integrated service delivery by facilitating collaboration between agriculture, health, education, and other sectors to address multifaceted rural development challenges.
6. *Targeting Marginal Groups:* ICT interventions target marginalized groups such as women, youth, and indigenous communities, providing them with equal opportunities for participation and empowerment.
7. *Creating Employment Opportunities:* ICT-enabled enterprises and digital platforms create new employment opportunities in rural areas, particularly in e-commerce, digital marketing, and rural service delivery.
8. *Maintenance of Land Holding Records:* ICTs facilitate the digitization and maintenance of land records, enabling transparent land tenure systems and efficient land management practices.
9. *Triggering Knowledge Revolution:* ICT interventions spark a knowledge revolution by democratizing access to information, fostering innovation, and promoting lifelong learning among rural communities.

Advantages of ICT

1. *Online Services:* ICT platforms offer a wide range of online services, including information dissemination, education and training, consultation, diagnosis, monitoring, and transaction processing.
2. *E-commerce:* ICT-enabled e-commerce platforms connect local producers directly to consumers, retailers, and suppliers, fostering transparent and efficient market linkages.
3. *Expert Consultation:* ICT platforms facilitate question-and-answer services, connecting farmers with domain experts to address specialized queries and technical challenges.
4. *Efficient Service Delivery:* ICT services delivered at the block and district levels enhance the efficiency and effectiveness of extension services, enabling comprehensive agricultural development.
5. *Timely Information Services:* ICTs provide up-to-date information on agricultural practices, market trends, weather forecasts, input supply, and credit availability, enhancing farmers' decision-making capabilities.
6. *Database Creation:* ICTs support the creation of local resource databases, site-specific information systems, and expert systems, enabling evidence-based planning and decision-making.

7. *Disease and Pest Management:* ICT platforms facilitate early warning systems for disease and pest outbreaks, enabling proactive interventions and mitigating agricultural risks.
8. *Land Records and Registration:* ICTs streamline land registration and record-keeping processes, reducing bureaucratic hurdles and enhancing land tenure security.
9. *Improving Marketing:* ICT interventions enhance the marketing of agricultural products, improving farmers' access to markets, value-added processing, and fair prices.
10. *Farm Business Management:* ICT platforms provide farmers with farm business and management information, enabling them to optimize resource allocation, investment decisions, and risk management.
11. *Cooperative Efficiency:* ICT-enabled communication networks and data-based technologies enhance the efficiency and productivity of agricultural cooperatives, promoting collective action and shared prosperity.
12. *Tele-Education:* ICT platforms offer tele-education services, providing farmers with access to online courses, tutorials, and agricultural training programs.
13. *Research-Extension Linkages:* ICT platforms facilitate the dissemination of research findings to extension workers and farmers, promoting evidence-based agricultural practices and innovations.

Scope of ICT in Rural Development

1. *Promoting Rural Business:* ICT interventions stimulate rural entrepreneurship, innovation, and enterprise development, unlocking new economic opportunities and diversifying rural livelihoods.
2. *Balanced Development:* ICT initiatives bridge the rural-urban divide by promoting equitable access to information, services, and opportunities, thereby fostering inclusive and sustainable development.
3. *Economic Diversification:* ICT interventions facilitate the diversification of economic activities in rural areas, promoting value addition, agribusiness development, and rural industrialization.
4. *Farm Information Dissemination:* ICT platforms serve as channels for disseminating farm information, agricultural advisories, and market intelligence, empowering farmers with knowledge and resources.

5. *E-governance:* ICT-enabled e-governance initiatives enhance service delivery, transparency, and accountability at the grassroots level, empowering rural communities and strengthening local governance structures.
6. *Natural Resource Management:* ICT interventions support sustainable management of natural resources by promoting precision agriculture, water conservation, land-use planning, and environmental monitoring.
7. *Enhancing Quality of Life:* ICT interventions improve the quality of life in rural areas by enhancing access to education, healthcare, financial services, and social amenities, thereby promoting human development and well-being.

Limitations of ICT in Rural Development

1. *Infrastructure Constraints:* Poor telecommunication infrastructure, inadequate power supply, and limited internet connectivity pose significant challenges to ICT deployment in rural areas.
2. *Service Providers:* Limited availability of internet service providers, technical expertise, and hardware engineers restricts access to ICT services and increases costs.
3. *Low Purchasing Power:* Low purchasing power among rural populations limits their ability to afford ICT devices, services, and subscriptions.
4. *Digital Divide:* Disparities in digital literacy and access to ICT services persist between rural and urban areas, exacerbating inequalities in information access and technology adoption.
5. *Media Coverage:* Limited coverage of rural issues by radio and television broadcasters hinders the dissemination of agricultural information and extension services to rural communities.
6. *Gender Sensitivity:* Gender disparities in ICT access, usage, and preferences impede women's participation in digital platforms and decision-making processes.
7. *Technical Challenges:* Slow connectivity, service disruptions, and system failures undermine the reliability and effectiveness of ICT services in rural areas.
8. *Awareness and Training:* Limited awareness and training on ICT technologies and applications constrain their uptake and utilization among rural communities and extension workers.

9. *Policy and Governance:* Inadequate policies, regulations, and strategies for ICT development and rural connectivity hinder the effective harnessing of ICT for rural development.
10. *Infrastructure Investment:* Insufficient investment in ICT infrastructure, human resources, and capacity-building initiatives constrains the scalability and sustainability of ICT interventions in rural areas.

The integration of Information and Communication Technologies (ICT) holds immense potential for revolutionizing agricultural extension and rural development. By addressing the multifaceted challenges of access, affordability, and capacity, ICT interventions can empower rural communities, enhance agricultural productivity, and promote sustainable livelihoods. However, concerted efforts are needed to overcome infrastructure constraints, digital divides, and governance issues to realize the full transformative impact of ICT in rural areas. Future research and policy initiatives should focus on leveraging ICT innovations to create inclusive, resilient, and technology-enabled rural societies.

23

Communication

Meaning and Nature of Communication

Communication Latin word 'Communis' meaning 'common'. Communication is the process of exchanging information, ideas, thoughts, feelings, or messages between individuals or groups. It is a fundamental aspect of human interaction and plays a crucial role in conveying meaning, building relationships, and facilitating understanding.

Nature of Communication

1. *Transactional*: Communication is a dynamic and interactive process involving both sending and receiving messages. It is not merely about transmitting information but also about interpreting and responding to messages, creating a continuous exchange.
2. *Multidimensional*: Communication occurs through various channels and forms, including verbal (spoken or written), nonverbal (gestures, facial expressions, body language), and visual (images, symbols, graphics). Different channels can convey different nuances of meaning.
3. *Purposeful*: Communication serves a purpose, whether it's to inform, persuade, entertain, or build relationships. The effectiveness of communication often depends on how well it achieves its intended purpose.
4. *Contextual*: Communication is influenced by the context in which it occurs, including the physical environment, cultural norms, social dynamics, and the relationship between the communicators. Understanding the context is essential for interpreting messages accurately.
5. *Transactional*: Communication involves an exchange of roles between senders and receivers. Both parties play active roles in the process, encoding and decoding messages, and adjusting their communication based on feedback received.

6. *Continuous Process*: Communication is an ongoing process that occurs continuously in various forms and stages. It involves a series of interconnected steps, including encoding, transmitting, receiving, decoding, and providing feedback.

Understanding the meaning and nature of communication is essential for effective interpersonal interactions, teamwork, leadership, and relationship building in both personal and professional contexts.

According to Leagans (1961), communication denotes the intricate process wherein two or more individuals exchange ideas, facts, emotions, or impressions in a manner conducive to achieving a shared comprehension of the conveyed message, its underlying intent, and the intended application of said message.

Nature of Communication

1. *Processual Nature*: Communication manifests as a dynamic process characterized by continuous interaction between communicators.
2. *Multiplicity of Means*: Communication utilizes a diverse array of channels, encompassing verbal, nonverbal, and visual modalities, to convey messages effectively.
3. *Interdependence*: Communication entails a mutual reliance between participants, wherein both the sender and receiver contribute to the exchange of information.
4. *Elemental Composition*: The communication process comprises discernible elements such as expression, interpretation, and response, which collectively shape the flow and comprehension of messages.
5. *Participant-Centric*: Communication occurs within the framework of participant interactions, wherein individuals engage in reciprocal dialogue to convey and receive messages.
6. *Multileveled Dynamics*: Communication operates across various hierarchical levels, ranging from intrapersonal exchanges within oneself to inter-organizational interactions spanning multiple entities.
7. *Fidelity Variability*: The fidelity or accuracy of message transmission varies across different communication contexts, influenced by factors such as noise, feedback mechanisms, and environmental conditions.

Purpose of Communication

1. *Persuasive Functionality*: The primary objective of communication often revolves around persuasion, wherein communicators endeavor to elicit a response or action from the recipient, appealing to their cognitive and emotional faculties.
2. *Informative Facilitation*: Communication serves as a means to disseminate information, catering to the intellectual faculties of the audience by imparting knowledge, facts, or insights.
3. *Entertainment Value*: Communication can also serve an entertainment function, offering recreational content designed to captivate and engage the audience's interest.
4. *Immediate and Delayed Gratification*: As posited by Schram (1964), communication may yield both immediate and delayed rewards, ranging from instantaneous feedback to long-term behavioral changes.
5. *Consummatory and Instrumental Dynamics*: Festinger (1957) delineates communication into consummatory, satisfying inherent needs, and instrumental, serving pragmatic objectives.
6. *Influence-Centric Orientation*: Berlo (1960) advocates for an influence-centered perspective on communication, emphasizing its overarching role in shaping human behavior and interactions from a behavioral standpoint.

Levels of Communication

1. *Intrapersonal Communication*: Communication occurring within the individual's psyche, encompassing internal dialogue, introspection, and self-reflection.
2. *Interpersonal Communication*: Direct, face-to-face interaction between two or more individuals, involving verbal and nonverbal exchanges within personal relationships or social encounters.
3. *Organizational Communication*: Communication within structured organizational frameworks, comprising:
   a) *Operational Communication System*: Pertaining to task-related activities and operational directives within the organization.
   b) *Regulatory Communication System*: Governing organizational protocols, rules, and regulations, ensuring compliance and order.

c) *Maintenance and Development of Communication System*: Focused on fostering feedback mechanisms and enhancing communication efficacy for organizational cohesion and growth.

4. *Inter-organizational Communication*: Communication channels facilitating interactions and collaborations between distinct organizational entities, aimed at achieving common objectives or addressing shared interests.

Communication Theories and Model

Communication models serve as symbolic representations of the structures, objects, or operations involved in the communication process. These models are characterized by either structural or process-oriented frameworks, elucidating the intricate dynamics underlying human interaction and information exchange. This paper aims to explore and elucidate the various communication models, with a particular focus on the linear model of communication.

Types of Communication Models

Linear Models of Communication

The linear model of communication delineates a one-way interaction paradigm, wherein feedback is notably absent. It represents the foundational framework of communication models, serving as the precursor to more complex models such as the transactional model. Within the linear model, communication unfolds as a unidirectional flow from sender to receiver, devoid of reciprocal feedback mechanisms. This model underscores the sender's role in conveying messages to the receiver without active engagement or response. Furthermore, it encapsulates the essence of a singular, linear process of communication devoid of iterative feedback loops or bidirectional exchanges. Example of linear models of communication is given below;

Communication Models	Year
Aristotle Model of Communication.	384-322BC
Lasswell's Model of Communication.	1948
Shannon-Weaver Model of Communication.	1949
Berlo's SMCR Model of Communication.	1960

Interactive Models of Communication: Dynamics and Implications

Interactive models of communication represent a dynamic paradigm characterized by bidirectional information exchange and feedback mechanisms. Unlike the linear model, which portrays a unidirectional flow of communication, interactive models emphasize the iterative nature of interpersonal interaction and the integral role of feedback in shaping communication dynamics. This paper aims to delve into the intricacies of interactive communication models, elucidating their significance in contemporary mediated and internet-based communication contexts.

Features of Interactive Models

1. *Two-Way Communication with Feedback:* The hallmark of interactive models is the facilitation of two-way communication between sender and receiver, allowing for reciprocal engagement and feedback exchange. This bidirectional flow enables participants to convey messages, interpret responses, and adjust their communication strategies accordingly, fostering a more dynamic and collaborative interactional process.
2. *Delayed and Indirect Feedback:* While interactive communication models encompass feedback mechanisms, the nature of feedback may vary in terms of speed and directness. Feedback in interactive models is often characterized by its gradual and indirect nature, resulting in slower response times and mediated feedback channels. Despite these temporal constraints, feedback remains essential for gauging message comprehension, clarifying misunderstandings, and refining communication outcomes.
3. *Potential for Linearity:* Although interactive models prioritize two-way communication, instances of linear communication may arise when receivers fail to provide timely responses or engage in reciprocal dialogue. In such scenarios, communication reverts to a linear format, with the sender predominating the discourse while awaiting feedback from passive receivers. This highlights the fluidity and contextual nature of interactive communication processes, which may oscillate between linear and interactive modalities based on participant engagement levels.
4. *Mediated and Internet-Based Communication:* The advent of digital technologies and internet-mediated platforms has significantly shaped contemporary communication landscapes, giving rise to novel forms of interactive communication. The interactive model of communication finds particular resonance in mediated contexts, including online forums, social media platforms, and virtual environments, where participants engage in asynchronous interactions characterized by delayed feedback and mediated communication channels.

In conclusion, interactive models of communication represent a dynamic and multifaceted framework that underscores the interactive nature of human interaction and information exchange. By facilitating bidirectional communication and feedback exchange, these models foster richer and more collaborative communication experiences, transcending the constraints of traditional linear communication paradigms. Moreover, the proliferation of mediated and internet-based communication platforms underscores the relevance and applicability of interactive communication models in contemporary digital environments, shaping the way individuals engage, interact, and communicate in an increasingly interconnected world. Example of interactive model of communication is given below;

Communication Models	Year
Osgood-Schramm Model of Communication	1954
Westley and Maclean Model of Communication	1957

Transactional Models of Communication: Dynamics and Distinctions

Transactional models of communication epitomize a nuanced framework characterized by bidirectional interaction and immediate feedback. Unlike linear or interactive models, transactional models emphasize the reciprocal nature of communication, wherein both sender and receiver exert influence on each other, shaping the ongoing dialogue. This paper aims to elucidate the intricacies of transactional communication models, delineating their distinctive features and highlighting their significance in fostering dynamic interpersonal exchanges.

Key Attributes of Transactional Models

1. *Two-Way Communication with Immediate Feedback:* At the core of transactional models lies the concept of bidirectional communication, wherein both parties engage in active discourse and feedback exchange. Unlike linear models, which lack feedback mechanisms, and interactive models, which may entail delayed feedback, transactional communication facilitates immediate and simultaneous feedback. This real-time exchange allows participants to respond promptly to incoming messages, fostering a dynamic and responsive communication process.

2. *Essential Role of Simultaneous Feedback:* Simultaneous feedback serves as a cornerstone of transactional communication models, enabling seamless interaction and mutual understanding between communicators. The immediacy of feedback ensures that responses are timely, direct, and contextually relevant, enhancing the efficacy and efficiency of

communication exchanges. In transactional models, both sender and receiver are actively engaged in the feedback loop, shaping the ongoing dialogue through continuous interaction.

3. *Direct and Rapid Feedback Mechanisms:* Feedback in transactional models is characterized by its directness and speed, with receivers compelled to provide instant responses to incoming messages. This direct feedback loop facilitates clear communication channels and minimizes the risk of misinterpretation or misunderstanding. The rapid exchange of feedback enables communicators to adapt their messages in real-time, fostering greater clarity and coherence in communication outcomes.

4. *Compulsory Receiver Engagement:* In transactional communication, receivers are inherently compelled to provide immediate feedback, as the dynamic nature of the interaction necessitates active participation from both parties. Unlike interactive models, where feedback may be optional or delayed, transactional communication mandates instant response from receivers, ensuring ongoing dialogue and mutual understanding.

Distinguishing Features from Interactive Models: While transactional and interactive models share similarities in their emphasis on bidirectional communication and feedback exchange, the key distinction lies in the nature of feedback. Transactional models prioritize direct and immediate feedback, whereas interactive models may involve delayed or indirect feedback channels. This distinction underscores the dynamic interplay between communicators in transactional models, wherein feedback serves as a catalyst for ongoing dialogue and mutual influence.

Transactional models of communication offer a comprehensive framework for understanding the dynamic nature of interpersonal interaction and information exchange. By emphasizing bidirectional communication and immediate feedback, these models facilitate responsive and adaptive communication processes, fostering mutual understanding and collaboration between communicators. The distinction between transactional and interactive models lies in the nature of feedback, highlighting the essential role of direct and rapid feedback mechanisms in shaping transactional communication dynamics. Thus, transactional models serve as invaluable tools for elucidating the complexities of human communication and fostering effective interpersonal exchanges in diverse contexts. Example of transactional model of communication is given below;

Communication Models	Year
Eugene White's Model of Communication	1960
Dance's Helical Model of Communication	1967
Barnlund's Transactional Model	1970

Different Communication Models

Aristotle Model *(384 to 322 B.C.) first basic persuasive model of communication*

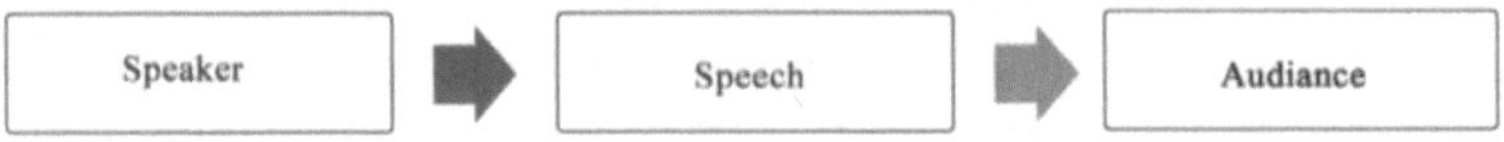

- Invention
- Organization
- Language
- Memory
- Delivery
- One way, linear

Lasswell's Communication Sequence *(1948, Rudimentary model of communication)*

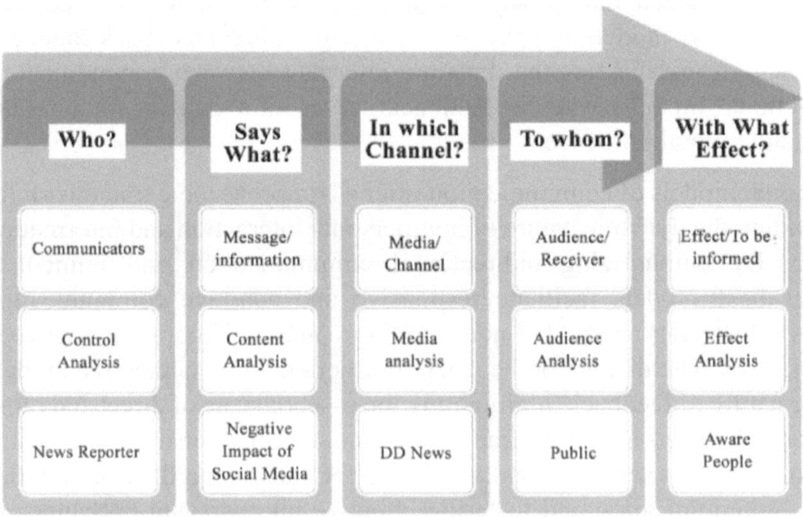

- Simple, linear, one-way model of communication
- Help to probe in to political communication, propaganda and political symbolism

Stimulus Response Model of Communication *(Interpersonal communication)*

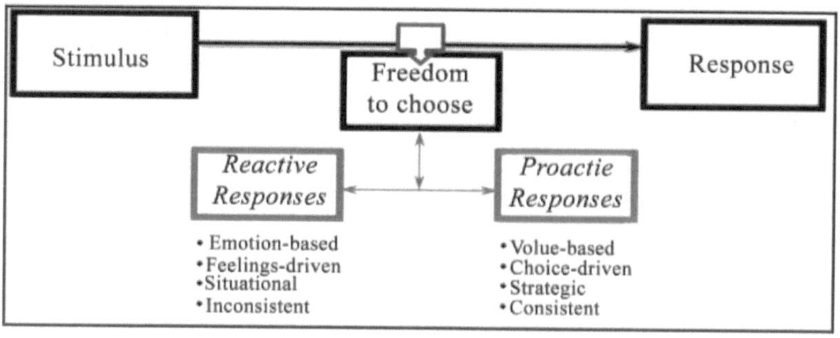

Schramm Model of Communication (1948)

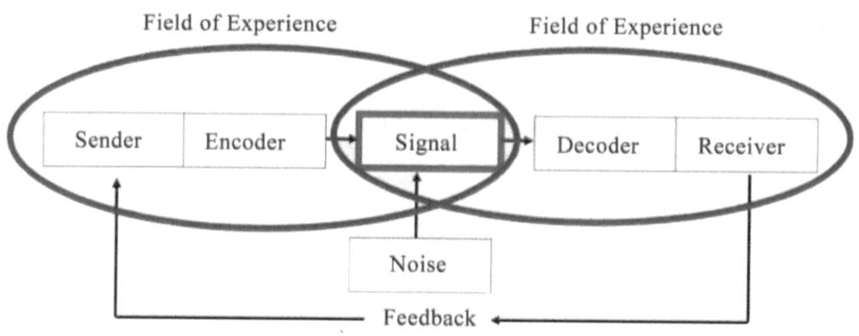

Schramm, W. (1995). Procedures and Effects of Mass Communicaton, in *Mass Media* and *Education*, ed Nelson G, Henny (Chicago University of Chicago Press.

- Accumulated field experience of both the source-encoder and receiver-decoder must be common for achieving maximum output
- Each person in communication process is both encoder and decoder
- Emphasis on organic nature of communication and continuous process

***Shannon & Weaver's Mathematical Theory of Communication** (Suitable for electronic media)*

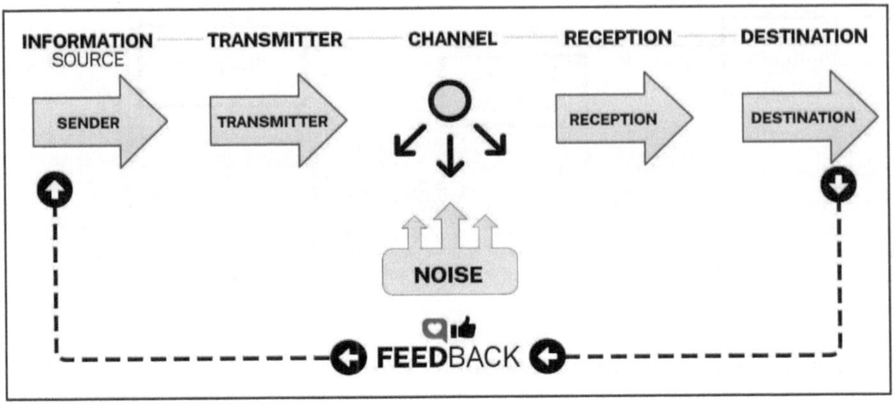

"Information Theory" in 1940

"Mathematical theory of Communication" in 1949

Three level of problems

- *Technical problems*- relate to accuracy of transference of information from sender to receiver.
- *Semantic problems*- related to interpretation of meaning by receiver as intended meaning of sender
- *Problems of influence or effectiveness*- concerned with success to which meaning conveyed to receiver lead to desired behaviour on his part.

Information: Relate to what could be said and not what is said.

Amount of information is measured in term of log 2 and unit of information is called BIT which is acronym to binary digit

Entropy: Information can be measured by entropy which is a measure of degree of randomness or choices or once degree of freedom of choice to select a message. The ratio of the actual to maximum entropy is called relative entropy

Redundancy: The amount of information that could be omitted or added in a noiseless channel, so that the message would still retain information or meaning.

Channel capacity: The amount of information that can be transmitted per unit of time.

Coupling: Any system that couples two other system is called a gate keeper.

Berlo's Model of Communication (1960)

The Berlo's Model ofCommunication was proposed by David K. Berlo in 1960. It's a simple and widely used model that helps in understanding the process of communication. Berlo's model emphasizes the interactional nature of communication and identifies key components involved in the communication process. The model consists of four main elements:

1. *Source*: The source refers to the person who initiates the communication process. It could be an individual, group, or organization that has a message to convey to the receiver.
2. *Message*: The message is the information, idea, or content that the source wants to communicate to the receiver. It can be verbal, non-verbal, written, or symbolic.
3. *Channel*: The channel is the medium through which the message is transmitted from the source to the receiver. It could be face-to-face communication, telephone, email, social media, or any other means of communication.
4. *Receiver*: The receiver is the person or audience for whom the message is intended. They decode and interpret the message sent by the source.

Additionally, Berlo's model also includes the concept of "encoding" and "decoding":

- *Encoding*: The process by which the source translates their thoughts or ideas into a form that can be understood by the receiver. This could involve choosing words, gestures, symbols, or other means to convey the message.
- *Decoding*: The process by which the receiver interprets and makes sense of the message received from the source. It involves understanding the message and assigning meaning to it based on one's knowledge, experiences, and cultural background.

Berlo's model suggests that effective communication occurs when there is clarity and understanding between the source and the receiver. It also acknowledges that communication is influenced by factors such as noise (anything that interferes with the transmission or reception of the message), feedback (responses from the receiver that indicate whether the message was understood), and context (the environment or situation in which communication takes place).

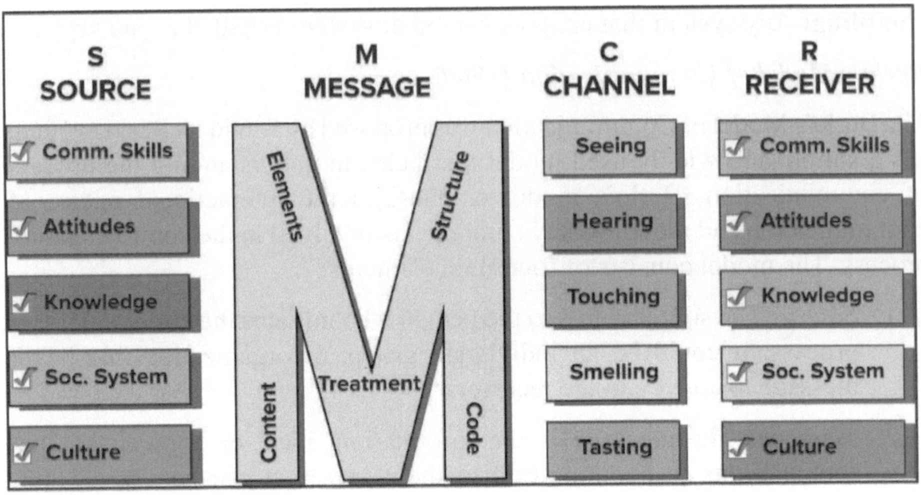

Leagans Model of Communication (1961)

- Communicator
- Message or Content
- Channel of communication
- Treatment of message
- Audience
- Audience response (concept of feedback)

There are several others that are commonly studied and referenced in communication theory. Some of the most prominent ones include:

Shannon-Weaver Model

Developed by Claude Shannon and Warren Weaver in 1949, this model is often referred to as the "Linear Model of Communication." It consists of a sender who encodes a message and transmits it through a channel to a receiver who decodes the message. It's a straightforward model focusing on the transmission of information.

Transactional Model

This model, proposed by Barnlund in 1970 and later expanded by scholars like Dance, is more complex than the linear models. It emphasizes the dynamic nature of communication, where both parties (sender and receiver) continuously exchange messages and roles. Communication is seen as an ongoing process influenced by feedback, context, and interpersonal factors.

Osgood-Schramm Model

Developed by Charles E. Osgood and Wilbur Schramm, this model builds upon the Shannon-Weaver model but adds the element of feedback and includes the concept of field of experience. It suggests that communication is influenced by the similarities and differences in the sender's and receiver's experiences.

Westley and MacLean Model

Proposed by Bruce Westley and Malcolm S. MacLean, this model emphasizes interpersonal communication within a social context. It highlights the role of feedback, where individuals adjust their communication based on the responses they receive from others, and also considers the influence of gatekeepers and opinion leaders in the communication process.

Linguistic theory of communication- spoken language

The linguistic theory of communication, particularly concerning spoken language, encompasses various aspects of language use and comprehension.

1. *Language as a System:* Linguistic theories often view language as a structured system comprising phonology (sounds), morphology (word formation), syntax (sentence structure), semantics (meaning), and pragmatics (language use in context). This systematic approach helps in understanding how language functions as a tool for communication.

2. *Speech Act Theory:* Proposed by philosophers like J.L. Austin and later developed by John Searle, speech act theory focuses on the performative aspect of language. According to this theory, when we speak, we don't just convey information; we also perform actions, such as making promises, giving commands, or asking questions. Understanding these speech acts is crucial in communication.

3. *Grice's Maxims:* Paul Grice proposed a set of conversational maxims that govern cooperative communication. These maxims include the maxims of quantity (providing the right amount of information), quality (being· truthful), relevance (staying on topic), and manner (being clear and avoiding ambiguity). Violations or flouting of these maxims can convey additional meaning or imply something beyond the literal interpretation of words.

4. *Pragmatics:* Pragmatics deals with the study of language in context. It explores how language users employ context, background knowledge, and shared assumptions to interpret meaning. Pragmatic factors, such as presuppositions, implicatures, and speech acts, play a crucial role in effective communication.

5. *Speech Production and Perception:* Linguistic theories also delve into the processes involved in speech production and perception. This includes studying how speakers plan and articulate their utterances and how listeners decode and interpret incoming speech signals. Models like Level's model of speech production and the Speech Perception Model provide insights into these processes.

The linguistic theory of communication, particularly regarding spoken language, encompasses various frameworks and concepts that help in understanding how language is used to convey meaning, perform actions, and achieve communicative goals in diverse social contexts. While J.P. Leagans might not be a prominent figure in this field, these broader theoretical frameworks offer valuable insights into the intersection of language and communication.

Barriers to Communication

Barriers to communication are obstacles or impediments that hinder the effective exchange of information between individuals or groups. These barriers can occur at any stage of the communication process and can significantly impact the clarity, accuracy, and understanding of the message being conveyed.

Types of barriers to communication

1. *Physical Barriers:* These barriers are related to the physical environment in which communication takes place. They include factors such as noise, poor lighting, uncomfortable seating arrangements, and distance between communicators. Physical barriers can disrupt the transmission of messages and make it difficult for individuals to hear, see, or focus on the communication.
2. *Psychological Barriers:* Psychological barriers stem from the attitudes, emotions, and perceptions of individuals involved in the communication process. These barriers may include preconceived notions, biases, stereotypes, fear, anxiety, and emotional states such as anger or stress. Psychological barriers can distort the interpretation of messages and impede effective communication.
3. *Semantic Barriers:* Semantic barriers arise from differences in language, vocabulary, and understanding of terminology between communicators. They can occur when words or phrases are interpreted differently, leading to misunderstandings or miscommunication. Semantic barriers are particularly common in cross-cultural communication and when technical jargon is used.

4. *Cultural Barriers:* Cultural barriers result from differences in cultural norms, values, beliefs, and communication styles between individuals or groups from different cultural backgrounds. These barriers can manifest in nonverbal communication cues, such as body language, gestures, and facial expressions, as well as in verbal communication patterns. Cultural barriers may lead to misinterpretation, offense, or the misalignment of expectations.
5. *Organizational Barriers:* Organizational barriers arise from structural or procedural aspects of the organization that impact communication. They include factors such as hierarchical structures, bureaucratic processes, lack of transparency, insufficient feedback mechanisms, and communication silos. Organizational barriers can inhibit the flow of information and impede collaboration and decision-making.
6. *Technological Barriers:* Technological barriers stem from the use of communication technologies that are unreliable, inaccessible, or incompatible with the needs of communicators. These barriers may include technical glitches, poor internet connectivity, incompatible software or hardware, and difficulties in navigating digital interfaces. Technological barriers can disrupt virtual communication channels and hinder the exchange of information.

Overcoming barriers to communication requires awareness, flexibility, and proactive efforts to address underlying issues. Strategies for overcoming barriers may include improving listening skills, enhancing cultural competence, clarifying messages, providing feedback, fostering open communication channels, and leveraging appropriate communication technologies.

24

Agricultural Journalism

Agricultural journalism is of recent origin in India. It is now gaining importance, particularly after the establishment of agricultural universities in India. Technical information needs to be provided to the farmers at the right time and in the right way to increase productivity. It is very much a development reporting activity in the context of agricultural development to inform, educate and motivate farmers to accept new ideas and agricultural practices in order to increase production. Agricultural journalism is a specific, focused and systematic mode of writing for the people engaged in farming as a business for their subsistence. Agricultural journalist collects the desired information and keeping in view the specific needs of the farmers disseminates it through various methods of communication such as print, radio, television and social media

Fundamentals of Agricultural Journalism

There are some fundamentals of agricultural journalism which have to be kept in mind so that the information flows smoothly and farmers are able to accept, evaluate, adopt and adapt. Following are the tips for agricultural journalists to make the information intelligible to the farmer.

Use of simple and plain language: The agricultural journalist should translate the technical terms in a simple language and follow inverted pyramid while writing for farm journalism.

Presentation of innovative ideas: People understand new ideas which are presented in a logical manner and it will help them get the best return for their investment.

To the point and brief message: Farmers are able to remember the information given to them point wise and in brief, directing towards advantages to be achieved. The message should not be verbose.

Use of visuals: Visuals make the message more interesting, catchy and more meaningful. Increasing use of visuals in a message makes it more intelligible and pleasing to the eyes.

Message is Recent, Reliable, Realistic and Relevant: Message which is framed by taking above consideration is highly acceptable to the farmers.

Steps for Effective Writing

Effective writing for farmers, Agricultural journalist should follow four important steps which are based on research findings for effective writing in the print media.

1. *Planning:* Think how to best achieve the objective of the message.
2. *Writing:* After proper planning, an agricultural journalist takes up the writing of the message for the media.
3. *Trimming:* It has been observed that many agricultural journalists tend to be verbose. Pruning of writing is equally important.
4. *Checking and re-checking:* We must be economical in the use of words notin giving information. The fewer words a sentence have the more its intelligibility

Rules to be followed by a journalist

1. Study the field and have an estimate of what type of articles is published in newspapers, magazines, books etc.
2. Look ideas
3. Develop the idea
4. Keep an unbiased mind
5. Write clearly and meaningfully
6. Remember the people to whom the articles are written
7. Write with a purpose
8. Write on timely subjects
9. Keep in touch with editors
10. Watch out for important events/occasions
11. Establish credibility, politeness and do not assume familiarity
12. Understand the communication process and the elements of communication.

ABC of Journalism

The ABC of Journalism refers to three fundamental principles that guide journalistic practice: Accuracy, Brevity, and Clarity.

1. *Accuracy:* Accuracy is paramount in journalism. It emphasizes the importance of reporting information truthfully, factually, and without bias. Journalists are expected to verify the accuracy of their sources, cross-check information from multiple sources if possible, and ensure that their reporting is based on reliable and credible sources. Accuracy helps maintain the trust of the audience and upholds the integrity of journalism as a profession.
2. *Brevity:* Brevity emphasizes the need for concise and succinct writing in journalism. It involves conveying information in a clear and concise manner, avoiding unnecessary details or verbosity. Brevity is essential in capturing the audience's attention, especially in today's fast-paced media environment where attention spans are limited. Journalists strive to communicate the essence of a story efficiently, making it easier for readers or viewers to understand and digest the information quickly.
3. *Clarity:* Clarity underscores the importance of clear and understandable communication in journalism. Journalists aim to present information in a way that is easily comprehensible to their audience. This involves using plain language, avoiding jargon or technical terms whenever possible, and structuring the content in a logical and coherent manner. Clarity ensures that the message is effectively communicated and minimizes the risk of misinterpretation or confusion among the audience.

By adhering to the ABC principles, journalists strive to uphold the highest standards of professionalism and ethics in their reporting, serving the public interest and facilitating informed decision-making.

25

Diffusion of Innovation

Diffusion of Innovation

Diffusion is the process in which an innovation is communicated through certain channels over time among the members of a social system. Element of Diffusion

Innovation

An innovation is an idea, practice or object that is perceived as new by an individual or other unit of adoption. The perceived newness of the idea for the individual determines his or her reaction to it. The "newness" aspect of an innovation may be expressed in terms of knowledge, persuasion or a decision to adopt.

Communication channels

A communication channel is the means by which messages get from one individual to another. The following classification of channels would help the communicator to use them appropriately:

(i) Interpersonal channels

(ii) Mass media channels

(iii) Cosmopolite channel

(iv) Localite channel

Time

Time is the most important element in the diffusion process. Time is an important aspect of any communication process. Time does not exist independently of events, but it is an aspect of every activity. The time dimension is involved in diffusion (i) in the innovation - decision process, (ii) in the innovativeness of an individual or other unit of adoption, and (iii) innovation's rate of adoption in a system.

Social system

It is a set of interrelated units that are engaged in joint problem solving to accomplish a common goal. The members or units of a social system may be individuals, informal groups, organisations and / or subsystems. The social system constitutes a boundary within which an innovation diffuses.

The Innovation Decision Process

The innovation decision process is the process through which an individual or other decision making unit passes from gaining initial knowledge of an innovation, to farming an attitude towards the innovation, to making a decision to adopt or reject, to implementation of the new idea and to confirmation of this decision. This process consists of a series of choices and action over time through an individual or a system evaluates a new idea and decides whether or not to incorporate the innovation into ongoing practices.

Innovation Decision Process consist of five stage-KPDIC

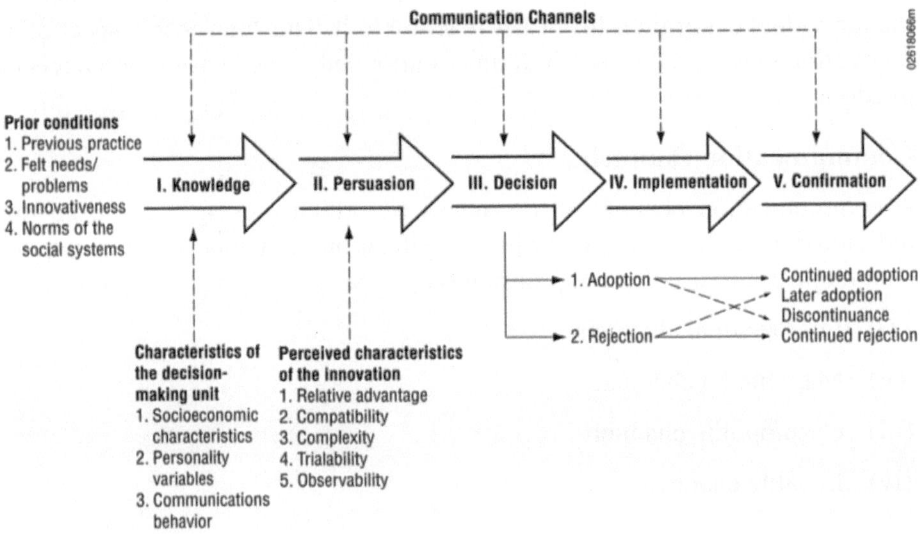

Knowledge stage

- Knowledge occur when an individual or other decision making unit is exposed to an innovation's existence and gain an understanding of how it function
- Selective exposure
- Selective perception

Type of Knowledge

- Awareness knowledge- information that an innovation exists
- How to Knowledge- consist of information necessary to use an innovation properly. The adopter must understand what quantity of an innovation to secure, how to use it correctly. To date few diffusion investigation deals with how to knowledge
- Principles Knowledge- consists of information dealing with the functioning principles underlying how an innovation works. It is usually possible to adopt an innovation without principles knowledge, but danger of misusing a new idea is greater and discontinuance result in paradigm shift in Innovation Decision Process

Persuasion stage

- Persuasion occur when an individual or other decision a making unit form a favourable or unfavourable attitude towards the innovation
- Attitude is a relatively enduring organization of an individual belief about an object that predispose his or her action
- Mental activity at knowledge stage was mainly **cognitive** (knowing)
- The main type of thinking at persuasion stage is **affective** (feeling)
- At persuasion stage individual become more **psychologically involved** with the innovation
- He or she actively seek information about the new idea, decide what messages he or she regard as credible, and decide how he or she interprete the information received
- Perceive attributes of innovation especially important at persuasion stage
    - Relative advantage, Compatibility and Complexity

Vicarious Trial: an individual may mentally apply a new idea to his or her present or anticipated future situation before deciding whether or not to try it. This vicarious trial involves the ability to think hypothetically and counterfactually and to project in to future. “Trial by others” provide a vicarious trial for an individual.

Cue-to-action: an event occurring at a time that crystallizes a favourable attitude into overt behavioural change. Some cue-to-action occurs naturally or may be created by a change agency.

Decision stage

- Decision takes place when an individual or other decision making unit engaged in activities that lead to a choice to adopt or reject the innovation.
- **Adoption-** is a decision to make full use of an innovation as the best course of action available
- **Rejection-** is a decision to not adopt an innovation
- **Discontinuance-** A decision to reject an innovation after having previously adopted it. Two types-
  - **Active rejection-** which consist of considering adoption of the innovation including even its trial but then deciding not to adopt it.
  - **Passive rejection-** which consist of never really considering the use of the innovation. Also called **non-adoption.**

Implementation stage

- Implementation occur when an individual or other decision making unit **put** a new idea in to use.
- Until implementation stage, the innovation decision processes has been astrictly mental exercise of thinking and deciding.
- Implementation involves overt behaviour change &fallow decision stagerather directly until logistic problems arise.
- Active information seeking usually takes place at the implementation stage
- Re-invention- is the degree to which an innovation is changed or modified byusers in the process of its adoption and implementation. Re-invention takes place at implementation stage of innovation decision process.

Confirmation stage

- Confirmation takes place when an individual or other decision making unit seek reinforcement of an innovation decision already made , but he or she may reverse this previous decision if exposed to conflicting messages about the innovation
- **Dissonance-** an uncomfortable state mind that an individual seeks to reduce oreliminate.

- **Discontinuance**-a decision to reject an innovation after having previously adopted it. Two types-
- **Replacement discontinuance**-a decision to reject an idea in order to adopt a better idea.
- **Disenchantment discontinuance**- a decision to reject an idea as a result of dissatisfaction with its performance.

Innovativeness

The degree to which an individual or other unit of adoption is relatively earlier in adopting new ideas than other member of social system. Innovativeness indicates overt behavioural change.

The S shaped Curve of adoption and Normality

- The time element of diffusion process allows us to classify adopter categories and draw diffusion curve.
- The adoptions of an innovation usually fallow a **normal, bell-shaped curve** when plotted over time on a **frequency basis**.
- If **cumulative number** of adopter is plotted, the result is an **S-shaped curve**.
- The part of diffusion curve from about **10 % adoption to 20 % adoption** is heart of diffusion process.
- The S shape curve is **innovation specific and system specific**, describing diffusion of a particular new idea among the member unit of a specific system.

Categorization of Adopters

Standardize adopter categories must decide on

- The number of adopter categories
- The proportion of the member of a system to include in each categories

The statistical methodIdeally categories should be

- Exhaustive, including all unit of study
- Mutually exclusive, by excluding a unit of study that appear in one categoryfrom also appearing in other category
- Derived from a single classificatory principle

Innovators

Venturesomeness is due to desire for rash, the daring and risky, and willing to accept an occasional setback, gate keeping role, not respected by others member of social system. The area lying to left of the mean time of adoption of an innovation minus two standard deviation includes innovators (2.5%)

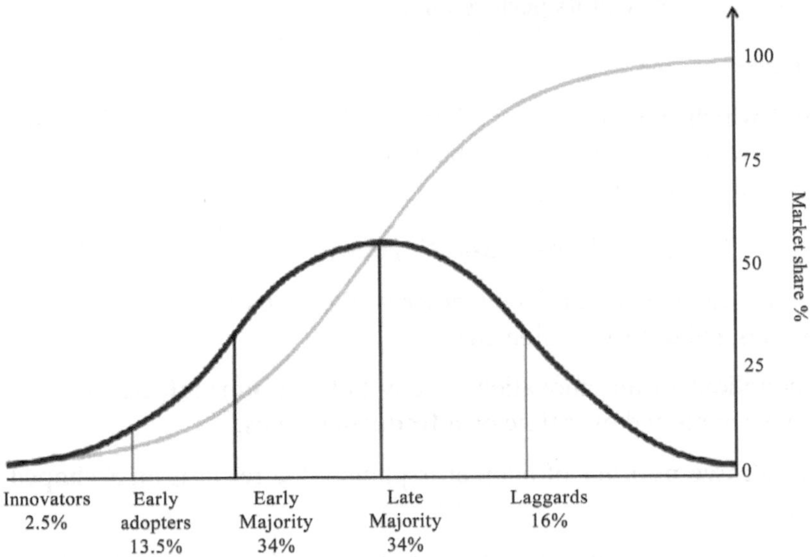

Early Adopters

Respectful, highest degree of opinion leadership, individual to check with ,local missionary, role model for other member of the social system, help to trigger the critical mass, make judicious innovation decision, subjective evaluation of innovation **(13.5 %).**

Early Majority

Deliberate, adopt new idea just before average member of a system, unique location, provide interconnectedness in a system, be not the first by which new is tried, nor the last to lay the old aside relate to them (34 %).

Late majority

Skeptical, adopt just after average member of a social system (34 %).

Laggards

Traditional, most localities, the point of reference of laggard is past, suspicious, limited resources (16 %).

Characteristics of Adopter Categories

a. Socioeconomic characteristics

- Earlier adopters are no different from later adopter in age.
- Earlier adopter has higher social status, large size units, more likely to literate,more years of formal education, greater degree of upward social mobility than do later adopters.

b. Personality variables

- Earlier adopter have greater degree of empathy (the ability of an individual to project himself or herself into the role of another person) than do later adopter.
- Earlier adopter s may be less dogmatic (Dogmatism- is the degree to which anindividual has a relatively closed belief system, that is, a set of belief whichare strongly held) than do later adopters.
- Earlier adopters have greater rationality (use of most effective means to reach a given ends) that do later adopters.
- Earlier adopters have more intelligence, favourable attitude towards science, favourable attitudes towards change, greater ability to deal with abstraction, better able to cope with uncertainty, higher aspiration than do later adopters.
- Earlier adopters are less fatalistic (Fatalism is the degree to which an individual perceive a lack of ability to control his or her future) than are later adopters.

c. Communication behavior

Earlier adopters have more social participation, highly interconnected through interpersonal network in their social system, more cosmopolites, more contact with change agent, greater exposure to mass media communication channel, greater exposure to interpersonal communication channel, seek information about innovation more actively, greater knowledge of innovation, higher degree ofopinion leadership than do later adopters.

Bibliography

Ahmed, M. R. (1991. Planning and Designing Social Forestry Project. In Ahmed, M. R. (ed.) Social Forestry and Community Development, pp. 109 - 116. FAO, FTPP.

Barrett, C. B., & Swinnen, J. F. M. (Eds.). (2008). Agricultural markets and rural poverty: empirical evidence from Tanzania. Springer Science & Business Media.

Chambers, R. (1983). Rural Development: Putting the Last First. Pearson Education India.

Claudio Zaki Dib (1988). Formal, Non-Formal and Informal Education: Concepts/Applicability. Cooperative Networks in Physics Education - ConferenceProceedings 173", American Institute of Physics, New York, pp. 300-315 retrieved from http://techne- dib.com.br/downloads/6.pdf

Davis, K., Nkonya, E., Kato, E., Mekonnen, D. A., Odendo, M., Miiro, R., & Nkuba, J. (2012). Impact of farmer field schools on agricultural productivity and poverty in East Africa. World Development, 40(2), 402-413.

De, D. and Jirli, B. (2010). A Handbook of Extension Education. Jodhpur, Agrobios (India). Development . Directorate of Extension, Govt. of India, New Delhi.

Dhama, O.P. and Bhatnagar, O.P. (1991). Education and Communication for Development. New Delhi, Oxford IBH Publishing Co.

Dimensions Of Agricultural Extension: (Aext191) (1+1). Principles, Philosophy Processes and Objectives of Extension Retrieved from http://eagri.org/eagri50/AEXT191/lec02.pdf

Dushi Guari. Notes on the Types of education: formal, informal, non-formal. Retrieved from http://www.preservearticles.com/notes/notes-on-the-types-of-education-formal-informal-non- formal/17995

Education from Wikipedia retrieved from https://en.wikipedia.org/wiki/Education Fundamental of Rural Sociology & Educational Psychology, (AEXT391) (2+0), lect1

Extension Program Planning and Evaluation. (2020 Jan. 10). retrieved from http://eagri.org/eagri50/AEXT392/lec02.html

Giarratano, J. C., & Riley, G. D. (2014). Expert systems: Principles and programming. Cengage Learning.

Gibson, D.V. & Smilor, W. (1991). Key Variables in Technology Transfer: A field – StudyBased on Empirical Analysis. Journal of Engineering and Technology Management, 8, p. 287-312.

Gibson, D.V., Rogers, E. & Wohlert, K. (1990). A Communication-based Model of TechnologyTransfer. Paper presented at the International Communication Association Meeting, Dublin,Ireland.

Govind, S., Tamilselvi, G. and Meenambigai, J. (2011). Extension Education and Rural Development, Jodhpur, Agrobios (India).

Gowda, K. N.(2016) Agricultural Extension Systems in India. retrieved from http://fliphtml5.com/bxhp/sxdi/basic

https://www.yourarticlelibrary.com/sociology/rural-leadership-meaning-and-characteristics-of-rural-leadership/34944 International Journal of Science Technology & Engineering, 1(5): 45-49.

Jackson, P. (1998). Introduction to expert systems. Pearson Education India.

Jaiswal, A. and Rai, S.K. (2019).Fundamentals of Agricultural Extension Education. Department of Extension Education College of Agriculture, Balaghat Jawaharlal Nehru Krishi VishwaVidyalaya Jabalpur 482 004 (M.P.),India pp. 1-62.

Joy, D. and Sreekumar, K. (2014). A Survey of Expert System in Agriculture.International Journal of Computer Science and Information Technologies, 5 (6): retrieved from https://newsmoor.com/3-types-of-communication-models-linear-interactive-transactional/

Lab to Land Programme. (2012, Jan 11). Retrieved from http://ecoursesonline.iasri.res.in/mod/page/view.php?id=15377

Lale, N. (1975). Rural Development. International Encyclopedia of Social Sciences.

Legans, J. P. (1961). Characteristics of Teaching and Learning in Extension Education, Extension Teaching Methods, The Communication Process and Programme Planning toMeet People's Needs in Extension Education in Community

Mishra, S. and Akankasha (2014). Expert Systems in Agriculture: An overview.

Mishra, S. N. (2007). Rural Development: Principles, Policies and Management. Sage Publications India.

Mondal, S. (2018). Communication Skills and Personality Development, Entrepreneurship Development and Business Communication. New Delhi,Kalyani publishers.

Mondal, S. (2019). Fundamentals of Agricultural Extension Education. Kalyanipublishers, New Delhi.

O.P. Dhama and O.P. Bhatnagar (2006). Education and Communication for Development,New Delhi: Oxford and IBH Publishing Co.

Principles of Extension Education, Dairy Extension Education.

Ramanathan, K. (1994). The polytrophic components of manufacturing technology.

Ray, G. L. (2013). Extension Communication and management. New Delhi. Kalayni publication.

Ray, G. L. (2017). Extension Communication and Management. New Delhi, Kalyani Publisher. Pp 227-248

Rogers, E.M. (1983). Diffusion of Innovations: Third Edition. The Free Press: A Division of Macmillan Publishing Co., Inc.

Sandhu A. S. (2003). Extension Programme Planning. New Delhi, Oxford and IBH Publishers.

Sasidhar, P.V. (2019 .(2.ceD(Programme Planning. retrieved from

Sharma, A. (2016). Education: Meaning, Definition, types of education and characteristics, Extension education: Meaning, definition and concept. Retrievedfrom https://www.slideshare.net/Arpita615/education-meaning-definition-types-of-education-and-characteristics extension-education-meaning-definition-and- concept

Sharma,O.P.(2012). Development perspective of extension education. Udaipur, Agrotech Publishing Academy.

Singh, R. (1987). A textbook of extension education. Ludhiana, Sahitya Kala Prakashan. Soil Health Card (2014, March 20)

Souder, W.E., Nashar, A.S. and Padmanathan, V. (1990). A guide to the best technology transfer practices. Journal of Technology Transfer, 15(1-2).

Sulaiman, V.R. and Sadamate, V.V. (2000). Privatising agricultural extension in India, Policy Paper 10. New Delhi: National Centre for Agricultural Economics and Policy Research. Technological Forecasting & Social Change, 46, 221-258.

Van den Ban, A.W. and Hawkins, H.S. (2002). Agricultural Extension, New Delhi, CBS https://newsmoor.com/3-types-of-communication-models-linear-interactive-transactional/

Winston, P. H. (1992). Artificial intelligence. Addison-Wesley.

World Bank. (2005). Agricultural extension and research: Achievements and problems in national systems. World Bank Publications.

Index